Detached-Eddy Simulation Series Methods for Compressor Internal Flow

Ruiyu Li · Limin Gao · Lei Zhao

Detached-Eddy Simulation Series Methods for Compressor Internal Flow

Method, Development and Application

Science Press
Beijing

Springer

Ruiyu Li
School of Power and Energy
Northwestern Polytechnical University
Xi'an, Shaanxi, China

Limin Gao
School of Power and Energy
Northwestern Polytechnical University
Xi'an, Shaanxi, China

Lei Zhao
School of Power and Energy
Northwestern Polytechnical University
Xi'an, Shaanxi, China

ISBN 978-981-96-0254-4 ISBN 978-981-96-0255-1 (eBook)
https://doi.org/10.1007/978-981-96-0255-1

Jointly published with Science Press
The print edition is not for sale in China mainland. Customers from China mainland please order the print book from: Science Press.

This Springer imprint is published by the registered company Springer Nature Singapore Pte Ltd.
The registered company address is: 152 Beach Road, #21-01/04 Gateway East, Singapore 189721, Singapore

Preface

Aviation compressors are one of the core components of aviation gas turbine engines, occupying 50 to 60% of the engine's length and 40 to 50% of its weight. Their performance critically influences the overall engine performance, cost, and safety. However, the turbulent flow within compressor blade passages is extremely complex due to the constraints of multiple walls and the effects of strong adverse pressure gradients.

Given the demand for an in-depth understanding of turbulence mechanisms in the development of high-performance aviation compressors, this book extends the high-fidelity Detached Eddy Simulation (DES) methods, which have been applied to external aerodynamics such as aircraft wings, to the more geometrically and flow-characteristically complex aviation compressors. From the perspective of engineering applications, this work integrates high-performance parallel algorithms, forming DES-based numerical methods tailored for compressor engineering design. Furthermore, it develops data analysis methods suitable for high-fidelity compressor turbulence "massive" data, supplemented by high spatial resolution experimental results, creating a systematic approach to understanding complex turbulent structures and flow mechanisms in compressors. This provides technical support for the refined design of high-performance aviation compressors.

This book encompasses the latest innovative results from our research group over the past decade, enabling readers to address practical engineering problems based on its content. Its main features are:

(1) **Integration of Theory and Practical Application**: It provides a comprehensive introduction to the theory and development of DES methods, the principles and characteristics of unsteady data analysis methods, and elaborates on the application of advanced numerical methods and data analysis techniques to solve practical physical problems, specifically focusing on compressors, a core engine component.
(2) **Meeting the Needs of Aviation Compressor Development**: Starting from the basic principles of the methods, it deeply analyzes the key scientific issues in applying high-fidelity numerical methods to simulate the complex internal

flow fields of compressors, introduces independently developed solutions, and explains the physical information that can be extracted through unsteady data analysis methods applied to compressor flow field analysis, targeting specific application areas.

This book covers high-fidelity numerical methods for large separation internal flows, methods for analyzing large volumes of unsteady flow data, and the application of these methods in understanding the turbulent structures in high-performance compressors. The specific contents are as follows: Chap. 1 summarizes the characteristics of complex flows within compressors and the challenging problems encountered in research. Chapters 2 and 3 explain the principles and equations of DES methods, addressing key scientific issues such as the difficulty in determining grid resolution, poor numerical stability, and the conflict between computational cost and accuracy when applying DES methods to complex compressor flow fields. Solutions proposed by the authors are also presented. Chapter 4 introduces various unsteady flow data analysis methods, including frequency analysis, vortex identification, and spatiotemporal decoupling of highly coupled flow fields, to fully extract the flow information embedded in the large volumes of high-fidelity data. Chapters 5–7 focus on real physical problems in the research field of high-performance compressors, using high spatial resolution experimental results as benchmarks to verify the applicability and value of DES methods in solving practical problems and uncovering the flow information contained in high-fidelity data.

The overall structure of this book is guided by the current bottleneck issues faced by compressors, providing a targeted analysis of the scientific challenges and solutions associated with the application of high-fidelity numerical methods to internal flow simulations within compressors. It elucidates efficient and high-precision DES-type methods tailored for the internal flow fields of high-load compressors and presents accompanying methodologies for analyzing unsteady, voluminous data. By adopting an unsteady research perspective, the book offers an in-depth understanding of compressor flow field structures and delineates technical approaches and their application outcomes for addressing practical problems. Through reading this book, researchers and engineers in the field of high-performance engine research will be introduced to new technologies and methods, while fluid mechanics researchers will gain insights into the challenging aspects of comprehending complex internal flow fields.

This work has been generously supported by the National Natural Science Foundation of China (grant numbers 52106053 and 92152301) and the National Science and Technology Major Project (No. J2019-II-0016-0037). The content of Chaps. 6 and 4 on SPOD is derived from the doctoral thesis of Mr. Shiyan Lin. We extend our sincere gratitude to Mr. Shiyan Lin for his contributions in authoring these sections, his profound understanding and insights into the gray area issues in Chap. 2, and for meticulously proofreading the manuscript. Additionally, we thank Mr. Guanzhong Ma for providing valuable materials for the modal decomposition section in Chap. 4 and for his thorough review of the content.

Due to the author's limited expertise, there may be inevitable omissions or shortcomings in this book. We warmly invite readers to provide constructive feedback and corrections.

Xi'an, China

Ruiyu Li
Limin Gao
Lei Zhao

Contents

Chapter 1
Introduction

1.1 Background

The compressor is the core component of aircraft engines and a crucial element determining their performance. Structurally, compressor components account for 50–60% of the total engine length and 40–50% of its weight. In terms of manufacturing and maintenance, compressors constitute 35–40% of the total manufacturing and 30% of maintenance costs for the engine. Furthermore, the aerodynamic performance of the compressor not only directly affects the overall engine characteristics (thrust, operational range) and thrust-to-weight ratio but also impacts flight safety, noise emissions, and pollutant emissions. Therefore, the performance of the compressor plays a critical role in aircraft engine development.

The modern aviation industry imposes increasingly higher demands on engine design, requiring engines not only to provide greater thrust but also to possess comprehensive attributes such as safety, reliability, cost-effectiveness, and environmental friendliness. Table 1.1 indicates that as performance requirements for aircraft engines continue to rise, the research and design of compressors face ever more stringent demands.

The history of compressor development demonstrates that compressor performance is greatly influenced by the internal flow characteristics. Consequently, in-depth research into the fine structure of its internal flow field and the mechanisms of energy loss is necessary to improve its aerodynamic performance and elevate compressor aerodynamic design to a new level. Due to the effects of rotation and strong curvature, secondary flows inevitably occur inside compressors in various forms, including passage vortices, corner vortices, leakage vortices, and wake vortices. Additionally, the interaction between stator and rotor components creates unique unsteady flow patterns within compressors. The internal flow in compressors is one of the most complex fluid dynamics encountered in practice. Its complexity arises not only from the intricate geometry of the flow path but also from the complex nature of the flow, primarily driven by unsteadiness. Research on unsteady, transonic,

R. Li et al., *Detached-Eddy Simulation Series Methods for Compressor Internal Flow*,
https://doi.org/10.1007/978-981-96-0255-1_1

Table 1.1 Trends in the development of aircraft jet engines

Generation	Engine type	Stage number	Pressure ratio	Average stage pressure ratio
Third	F100	10	7.8	1.228
	F110	9	9.7	1.287
	RD33	9	7	1.241
	WS10	9	9.7	1.287
Fourth	F119	6	8	1.414
	M88-2	6	6.45	1.364
	EJ200	5	6.2	1.440
	WS15	6	6	1.348
Fifth	AXXX	4	6	1.565

and three-dimensional flow fields constitutes a major research direction in modern compressor studies. Thus, the investigation of internal compressor flow fields is a research topic of both theoretical and practical significance.

Theoretical analysis, experimental measurements, and numerical simulations are three important approaches to study compressors. Theoretical analysis is the earliest and most fundamental method used to investigate internal compressor flow. It can delve into the fundamental laws governing the problems and provide the most clear guidance. Lift theory and cascade theory form the basis for all aerodynamic research on compressors. However, the flow control equations, solution domains, and boundary conditions of compressors are highly complex. Theoretical analysis cannot consider all factors and their interactions, and it cannot provide practical theoretical solutions to meet industrial development needs.

Experimental measurement is another important means of studying fluid flow mechanisms, considering various factors that influence flow. Results obtained from experiments are objective and reliable, serving as a direct test of the correctness of theory. Numerous systematic cascade experiments have been conducted, leading to the establishment of general principles for compressor design. Early compressor design was largely based on extensive experimental data, guided by theory. However, compressors are the most structurally complex fluid machinery, and experiments are costly, time-consuming, and often lack adequate measurement techniques for capturing the intricate flow structures, including those inside the boundary layer. Even with advanced measurement tools such as Laser Doppler Velocimetry (LDV), Particle Image Velocimetry (PIV), and Pressure-Sensitive Paint (PSP), precise measurements of flow inside the boundary layer and the unsteady flow on all scales remain challenging. Additionally, experiments may not always replicate similar conditions, as they are susceptible to unavoidable external disturbances. The limitations in our understanding of flow mechanisms also make it difficult to obtain completely accurate results.

The advent of computers has made numerical solutions of partial differential equations possible, and computational fluid dynamics (CFD) has provided a new way to understand internal compressor flow. Compared to experimental measurements, numerical methods offer advantages such as lower cost, shorter timeframes, and rich data. As computer and computational technology continue to advance, CFD technology has garnered increasing attention as a simpler extension of theoretical analysis and experimental measurement. In many cases, CFD has been able to accurately simulate real flow, serving as a "numerical wind tunnel" and replacing numerous wind tunnel experiments. It has become a powerful and effective tool for studying the complex internal flow of compressors, becoming one of the key tools in compressor design.

In recent years, the trend in the design method for high-performance compressors has shifted from early "simplified design" methods to modern "fine design" approaches. This transition emphasizes not only achieving macroscopic aerodynamic indicators (pressure ratio, efficiency, etc.) but also effective control of internal flow details (shock waves, separation flow, unsteady flow, etc.) and their influencing factors under adverse conditions such as near-stall and near-surge. This shift in design philosophy places higher demands on numerical simulations, necessitating the high-fidelity prediction of complex flow phenomena, such as shock waves, flow separation, transition, and interactions under these adverse conditions. Currently, the widely used numerical methods based on Reynolds-averaged equations are limited by turbulence models and cannot effectively handle the flow under such adverse conditions. Thus, there is an urgent need to develop more advanced numerical methods that can finely reveal flow characteristics and mechanisms within blade passages, providing a deeper understanding of the sources and influencing factors of flow losses. This is a key foundational research for improving compressor aerodynamic design and, in turn, enhancing the overall performance of aircraft engines.

1.2 Compressor Internal Flow Characteristics and Current Status

1.2.1 Compressor Internal Flow Characteristics

The aerodynamic performance of compressors is heavily dependent on the complex, constrained turbulent flow created by adjacent blades, the hub, and the casing. This complexity arises not only from the intricate geometry of the flow path but also from the complex nature of the flow itself. Due to the complex blade passage shape, the centrifugal force generated by rotation, and Coriolis forces, the internal flow of compressors is essentially three-dimensional and unsteady. Additionally, the presence of fluid viscosity and the interaction between rotating and stationary blade rows make the internal flow even more structurally complex in time and space.

The complexity manifests in several ways:

- Under adverse pressure gradients, the boundary layer on the blade surface rapidly develops and is prone to separation due to the challenging pressure gradients it cannot overcome.
- The combined effect of pressure differences in the spanwise direction and the endwall boundary layer results in the inevitable occurrence of secondary flows and passage vortices.
- The narrow and curved passages created by the blades and endwall not only accelerate boundary layer development and separation but also lead to complex flow structures, including corner flow, corner vortices, gap flow, and leakage vortices, formed due to mutual interference.
- Effects such as centrifugal forces from high-speed rotation, shock waves from supersonic flow, shock/boundary layer interactions, and boundary layer thickness stacking in the circumferential direction cannot be ignored.
- Interactions between the rotor and stator, including wake-blade interactions and temporal effects, contribute to the complexity.
- Unsteady aerodynamic forces result in flutter phenomena, and large-scale unsteadiness, such as rotating stall and surge, can occur under high-load conditions.

As shown in Fig. 1.1, the spatiotemporal scales of flow structures within the blade passage of aircraft engine compressors span five to six orders of magnitude, from the smallest turbulent fluctuations to the largest transient flow structures (occurring during dynamic processes such as startup, shutdown, throttling, inlet distortion, etc.). Consequently, the internal flow within compressor blade passages is considered one of the most complex flow phenomena encountered in practice.

High loading capability is one of the key objectives in designing high-performance compressors. From an aerodynamic perspective, increasing the compressor's loading capability can be achieved by increasing the blade's camber angle and raising the blade tip speed, as shown in Fig. 1.2. However, large camber angles and high tip speeds bring about numerous aerodynamic challenges, including extensive separation and shock/boundary layer interaction problems in the flow field. The most severe issues occur in the three-dimensional corner regions at the root of the compressor blade and in the clearance flow region at the blade tip, particularly in the areas of three-dimensional corner separation and tip clearance flow issues.

Specifically, within the three-dimensional corner region formed by the blade root and the lower endwall (hub) (Fig. 1.3), under the influence of high curvature and strong adverse pressure gradients, the boundary layers on the blade surface and endwall develop into a "three-dimensional corner separation" structure composed of multiple turbulent quasi-ordered structures. In the clearance gap at the blade tip and upper endwall (casing) (Fig. 1.4), phenomena such as tip clearance leakage flow, casing boundary layers, shock waves, and relative motion-induced scraping interactions create a complex turbulent structure characterized by "shock waves, boundary layers, and leakage vortices" mutual interference.

Numerous studies indicate that as compressor loading increases, more than 50% of flow losses and locations of flow instability concentrate in the blade's ends (tip

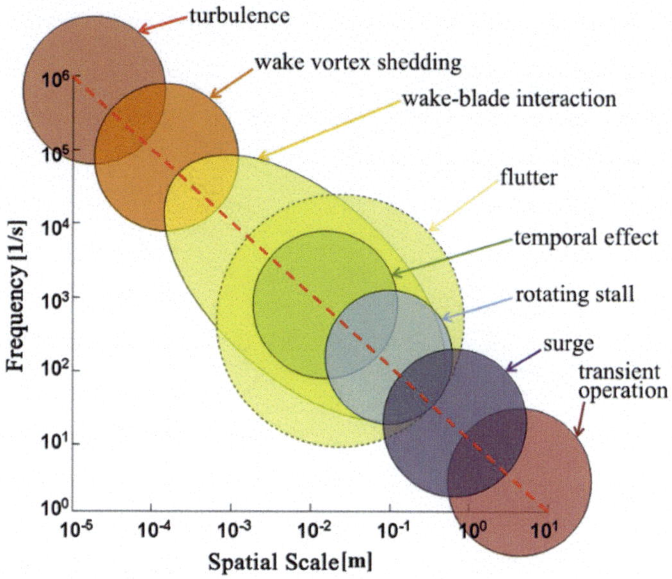

Fig. 1.1 Temporal-spatial spectral distribution of unsteady flow in compressors [1]. *Source* LaGraff et al. [1]

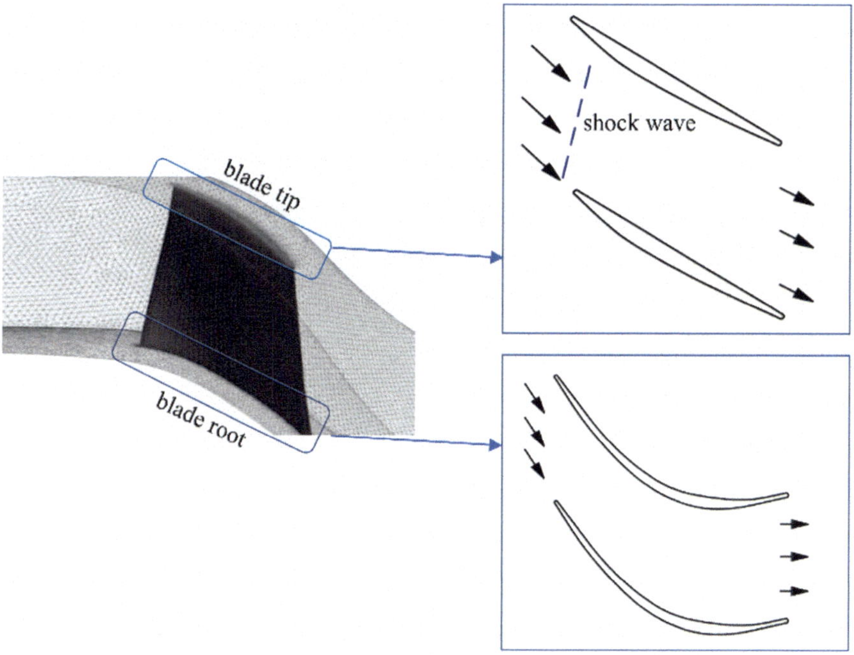

Fig. 1.2 Schematic of the characteristics of the hub and tip regions in compressors

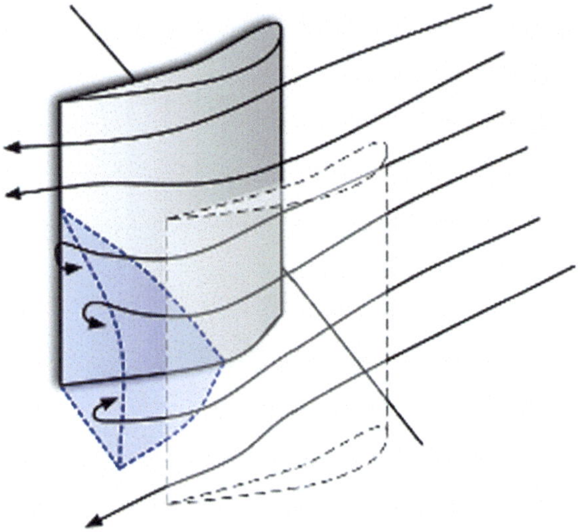

Fig. 1.3 Schematic of three-dimensional corner separation at the blade root region [2]

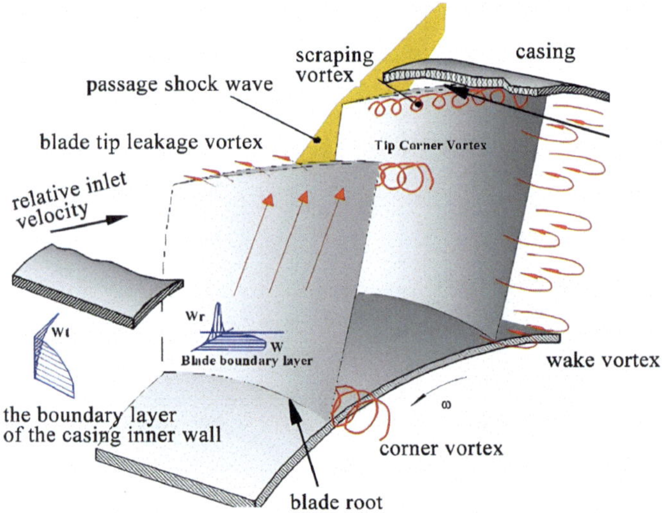

Fig. 1.4 Schematic of complex flow in the blade tip region

and root). Furthermore, these issues have direct or indirect effects on aerodynamic stability (Fig. 1.5). Currently, the study of large separation at the blade root and shock/vortex interference at the blade tip are hot topics in compressor research.

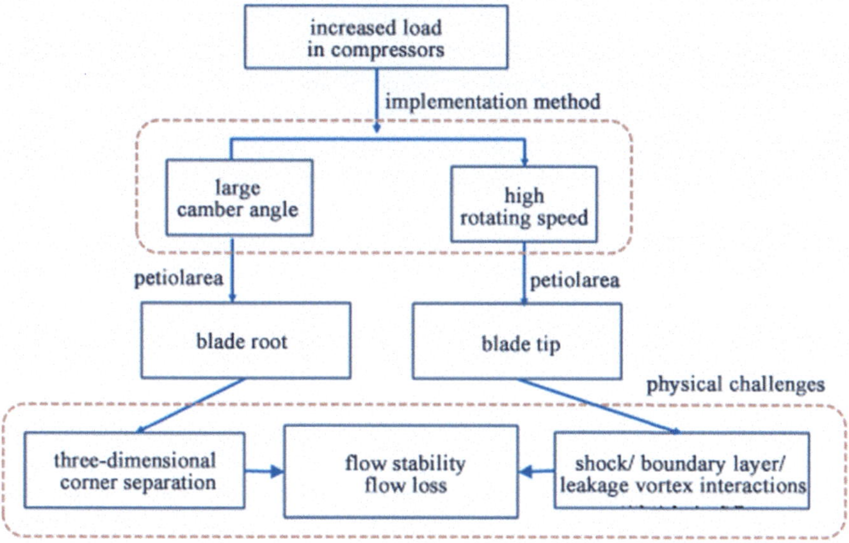

Fig. 1.5 Physical challenges arising from increased load in compressors

1.2.2 Understanding Compressor Internal Flow

(1) **Blade Tip Flow**

In compressors, there is inevitably a clearance gap at the blade tip to allow for relative motion between the rotor and the casing. Due to pressure differences between the pressure and suction sides of the blades, some airflow leaks over the blade tip, forming leakage flow. This flow interacts with the casing boundary layer, the mainstream flow, and other secondary flows, causing significant flow losses [3]. Although the gap size is less than 1.0% of the blade height [4], its impact area occupies nearly 20% of the blade passage [5]. The losses it causes typically account for 20–40% of the total losses [6]. Since the 1950s, the flow near the blade tip has been a hot topic of interest to researchers.

Early research on blade tip leakage flow aimed to establish simple flow models, such as Rains' model based on the inviscid assumption [7], the Lakshminarayana model [8], and the Chen model that does not consider blade thickness [9]. These models estimated factors affecting tip leakage flow, leakage flow rates, and flow losses [7–13], providing references for compressor design and contributing to an initial understanding of the nature of tip leakage flow. However, these simplified models, which incorporated numerous empirical assumptions to account for various influencing factors, lack strict accuracy and have limited applicability. Therefore, purely theoretical model-based research cannot deeply investigate the mechanisms and patterns of tip leakage flow.

For more detailed studies of the tip region flow, researchers turned to the development of RANS (Reynolds-Averaged Navier–Stokes) numerical methods. By examining steady-state results, researchers investigated the impact of different tip clearance structures on flow loss and focused on studying macroscopic parameters and summarizing their influence patterns. Subsequently, numerous studies have shown that tip leakage flow has a significant impact on inducing compressor instability [14–16]. This drove research into the regularities associated with tip clearance structures and instability, including the impact of tip clearance structures on stability [16], flow characteristics under instability conditions [15], and analyzing changes in flow field structure before and after instability occurs [14, 17]. The goal was to achieve control over instability. While analyzing macroscopic parameters and regularities is helpful for understanding specific issues in certain models, it is often limited in generality without a fundamental understanding of the physical mechanisms involved.

Since 2000, research has increasingly indicated that under harsh conditions like near stall, tip clearance flow exhibits strong unsteadiness [4–18]. Studies conducted within a steady framework, due to the averaging process, may lose important flow details, such as the formation of leakage vortices and the breakdown of vortex systems [15–21]. Therefore, the academic community has shifted its focus to non-steady frameworks to study the unsteady pulsations, development, and evolution of tip clearance leakage flow, aiming to understand its impact on stability. Currently, it is generally believed that the low-frequency, large-scale "modal waves" and high-frequency, small-scale "high-frequency disturbance waves" in the tip region are one of the main causes of inducing compressor stall [21]. However, there are different interpretations of the reasons behind the induction of these non-steady disturbances. Some argue that the unsteady oscillations in tip clearance flow result from the breakdown of tip clearance vortices [15–17], while others suggest that it is due to the enhanced interaction between tip clearance flow and the mainstream flow, regardless of external excitation and vortex breakdown [4, 18]. Furthermore, at high loads, the high tip speed leads to strong shock waves at the blade tip. The presence of shock waves introduces new non-steady disturbances associated with shock-wave/boundary-layer interactions and leakage vortices (Fig. 1.6), making the understanding of the root causes of compressor instability even more complex.

(2) **Blade Root Flow**

In the presence of adverse pressure gradients within compressors, the boundary layer attached to the hub endwall and the blade suction surface (the leeward side) gradually thickens in the blade passage. In the area between the blade root on the suction surface and the hub endwall, the boundary layers of the suction surface and the endwall interact, forming a region of low-energy fluid aggregation known as the corner boundary layer. When low-energy fluid cannot overcome the adverse pressure gradient, flow separation occurs in the corner region formed by the blade root and the endwall, as shown in Fig. 1.7. The separation phenomenon occurring within the corner region formed by the suction surface of compressor blades and the endwall is referred to as "corner separation." Due to the development of this structure in three directions, namely streamwise, spanwise, and tangential, it is also termed

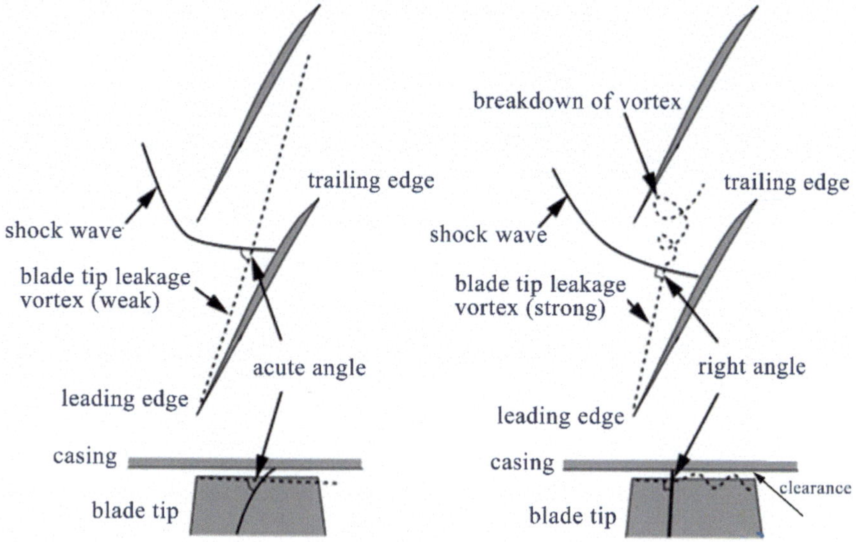

Fig. 1.6 Schematic of wave/vortex interactions in the blade tip region [17]

"three-dimensional corner separation." This flow phenomenon represents one of the most significant influences on the performance of high-loaded compressor blade hub regions.

Corner separation is a common phenomenon in compressor internal flow fields [22]. Under relatively stable operating conditions, corner separation only occurs in a small area near the suction surface of the blade and does not significantly impact compressor performance. Initially, it did not receive much attention from researchers. It was only in the 1980s and 1990s, with the demand for high-loading designs, that blade profiles with large camber angles (even exceeding 60°) began to

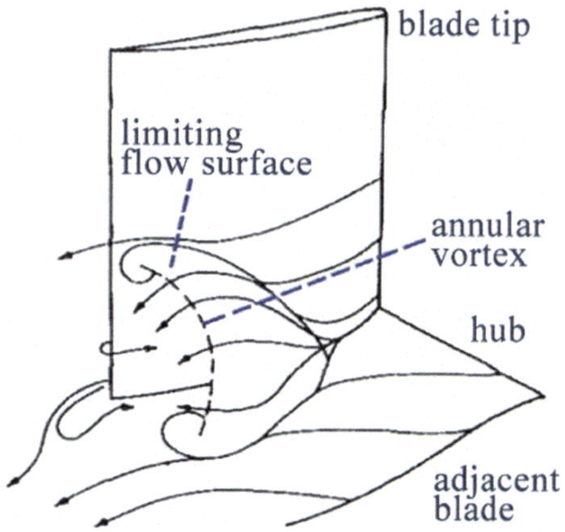

Fig. 1.7 Schematic of three-dimensional corner separation at the blade root region [24]

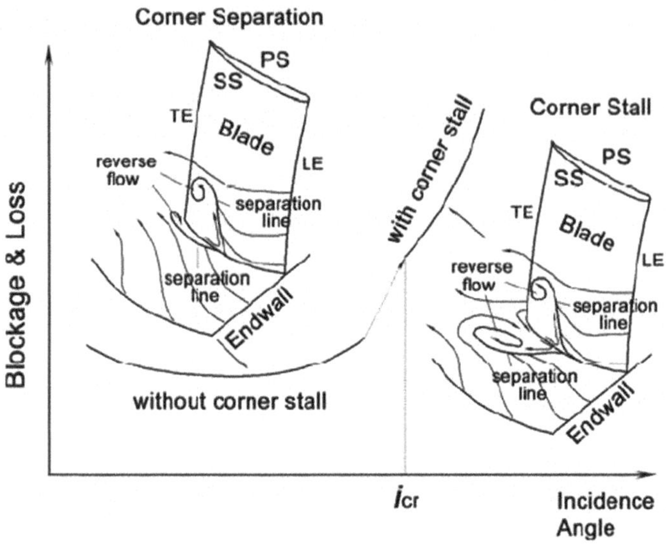

Fig. 1.8 Schematic of corner separation and corner stall topology and aerodynamic performance [25]

be used at the blade root. Under the influence of high-curvature blade profiles, the three-dimensional effects of corner separation flow were greatly enhanced, leading to a substantial increase in the extent of corner separation, referred to as "large separation." At this point, the losses sharply increased, and the pressure-raising capacity decreased. Lei et al. [23] suggested differentiating between "corner separation" and "corner stall" (Fig. 1.8). To avoid confusion and simplify terminology, this book refers to "corner separation" or "three-dimensional corner separation" to denote the significant flow separation phenomenon (including corner separation and corner stall) that affects aerodynamic performance under high loads.

The understanding of corner separation in the blade root region is similar to that of tip leakage flow. Early researchers constructed spatial three-dimensional topological structures based on friction lines on the walls. One of the most representative works is the topology of corner separation proposed by Schulz and Gallus in 1990 [24]. Although it did not provide detailed flow field structures inside the separation region, it clearly depicted the basic characteristics of the flow: corner separation initiates near the leading edge of the blade on the suction surface and the endwall, and once initiated, it severely obstructs the entire flow field, gradually extending into the spanwise and tangential directions, forming a ring-shaped recirculation zone.

Around the early twenty-first century, with advancements in numerical simulation methods and experimental measurements, it became possible to gain insights into some details of the spatial flow field within the corner region by combining numerical simulations of three-dimensional flow fields with experimental measurements. Researchers began to realize that corner separation flow is an extremely complex three-dimensional flow. This complexity is manifested in two aspects: the complexity of the flow structure itself and the complexity of the inducing factors.

Regarding the flow structure of corner separation, in 1999, Hah and Loellbach [26] conducted numerical simulations of a transonic compressor rotor using the RANS method and proposed a new three-dimensional spatial topology, as shown in Fig. 1.9. This topology not only demonstrates the three-dimensional development characteristics of corner separation but also presents the characteristics of the vortex structures within the separation region. Specifically, within the spatial recirculation zone bounded by the extreme flowlines of the blade suction surface and endwall, there are two main vortex structures. One is located approximately 80% chord length near the suction surface of the blade, and the other is near the trailing edge. These two vortices rotate in opposite directions and develop tangentially, eventually merging outside the boundary layer (as shown in Fig. 1.9). Hah suggested that corner separation is primarily dominated by a tornado-like vortex structure, and the two vortices mentioned earlier are two legs of the tornado-like vortex structure attached to the hub. Lewin [27] suggested that these are independent vortex structures that interact and are not derived from the same vortex structure. He provided a more intricate topology of corner separation, as shown in Fig. 1.10.

The mechanism behind the formation of corner separation is currently understood in terms of three macroscopic factors [22, 26, 28]: (1) strong adverse pressure gradients, (2) secondary flows within the passage, and (3) mixing between the endwall and blade surface boundary layers. However, a mature understanding of the reasons secondary flows induce corner separation, the development and evolution of turbulent pre-swirl structures, and their interplay is still lacking. These aspects are critical for the fine design of high-performance compressors.

With the increasing accuracy of numerical simulations (achieving higher spatial resolution through increased grid density), researchers have made deeper progress in

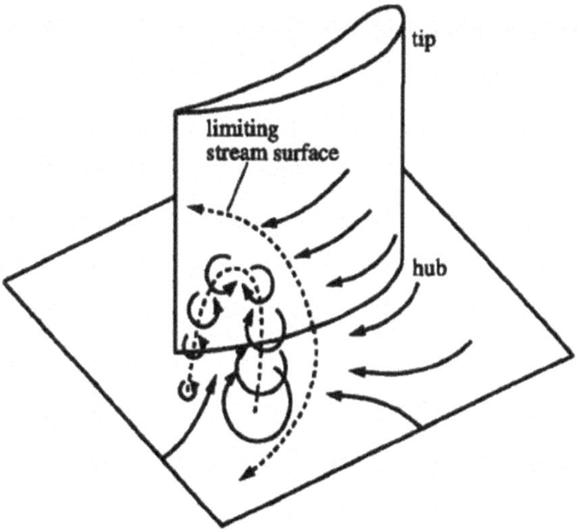

Fig. 1.9 Hah's corner separation topology [26] (based on RANS results)

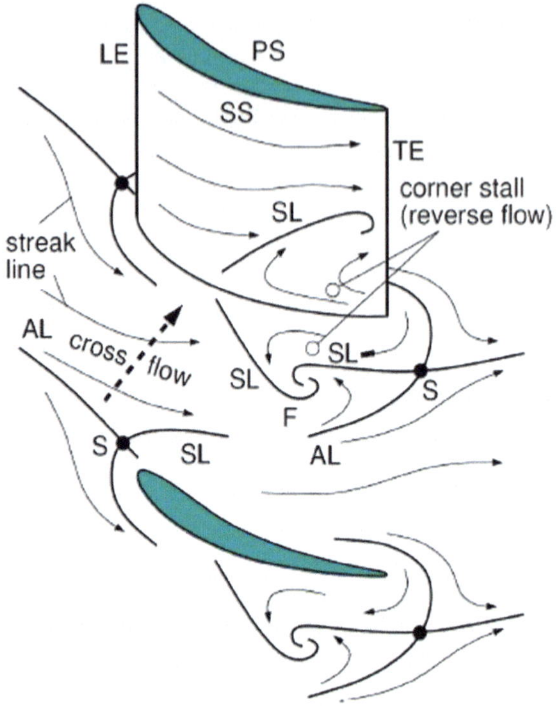

Fig. 1.10 Lewin's corner separation topology [27] (based on RANS results)

understanding the corner region flow. It has been recognized that three-dimensional corner separation is not solely the outcome of a single vortex structure but rather the result of the intermixing and evolution of multiple vortex systems. These vortex structures interact, blend, and couple with each other, exacerbating the complexity of understanding the mechanisms governing their mutual interference. Many studies on the spatial topological structures of vortex systems [29, 30] have proposed multiple models for corner vortices (as shown in Fig. 1.11). However, there is no consensus on the locations, influences, and dependencies of these vortices. For example, Su [31] and Xia [32] both captured the formation of "corner vortex" structures, but Su suggested that these structures originated outside the blade passage and resulted from the development of separated flow, while Xia observed them within the passage and believed they formed from the breakup of horseshoe vortices. Gbadebo [33] and Gao [34] found a close connection between horseshoe vortices and three-dimensional corner separation structures, but the conclusions regarding the relationship between these structures are quite uncertain. Chen [35] studied the flow field in a specific compressor blade passage and found that three-dimensional corner separation was induced when two branches of the horseshoe vortex met. However, in Liu's work [36], even though the two branches of the horseshoe vortex did not meet, severe corner separation still occurred.

It is evident that corner separation results from the highly coupled actions of multiple vortex structures, and these vortex structures induce, interfere with, and

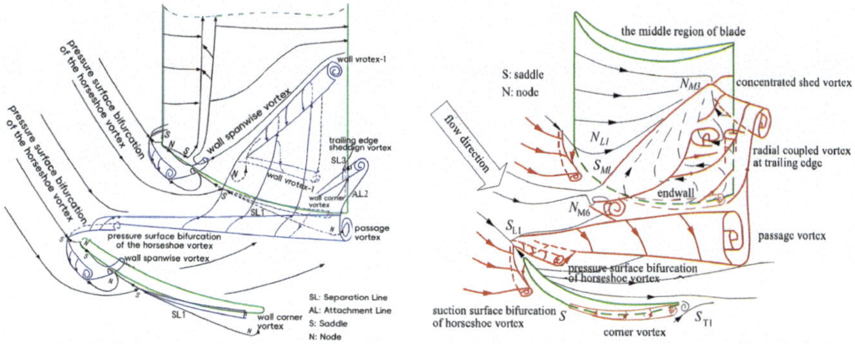

Fig. 1.11 Rotating vortex models for compressor corners (left: Zhang [29], right: Zhao [30])

couple with each other. These interactions intensify the complexity of the corner region flow field, making it an open problem.

Fundamentally, the generation, mixing, and evolution of vortex structures inevitably introduce strong unsteadiness to the local flow field [1]. Using SPIV (Stereo Particle Image Velocimetry) and laser Doppler experimental techniques to measure the corner region flow has confirmed this point [37, 38]. Viewing the flow from a steady perspective can lead to the loss of essential flow details. For example, Furukawa [16, 19] conducted numerical simulations of a near-stall condition in a subsonic axial compressor rotor using both steady and unsteady approaches. In the steady results, there was an anomaly in which the leakage flow did not roll up into leakage vortices. However, from an unsteady perspective, it was evident that the leakage vortices were breaking down. Clearly, under extreme conditions in high-load compressors, the flow field with large-scale separation exhibits non-steady characteristics. Understanding the underlying mechanisms of these flow phenomena requires studying them within a non-steady framework.

In summary, the internal flow of compressors is characterized by the presence of multiple complex vortex structures that are highly coupled and exhibit unsteadiness. Understanding the physical phenomena in the compressor's root region, such as corner separation and tip leakage flow, depends to a large extent on the development of research methods. As shown in Fig. 1.12, inviscid-based numerical simulations provided initial insights into compressor flow fields. With the development of RANS numerical simulation methods and increased computational power, researchers gained preliminary knowledge of the complex flow field structures within compressors and began to recognize various complex vortex structures. Non-steady numerical simulation techniques revealed the non-steady nature of tip clearance flow and brought to light some unsteady aspects that were previously hidden in steady-state studies. Consequently, further advancements in high-fidelity numerical and experimental methods will continue to drive our understanding of the mechanisms of compressor internal flow.

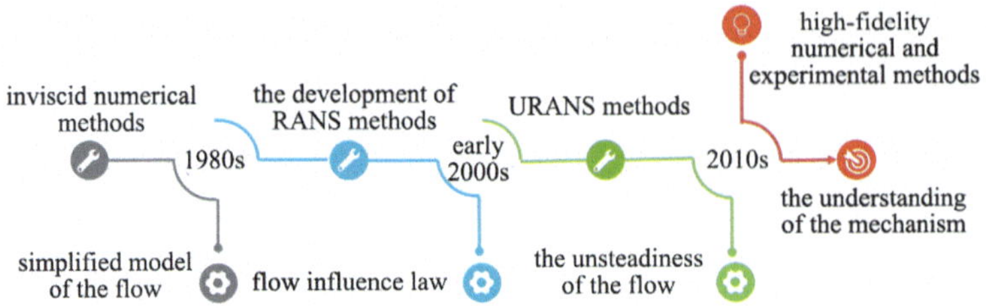

Fig. 1.12 Development of understanding of internal flow in compressors based on numerical simulations

1.2.3 Challenges in Numerical Simulation of Internal Compressor Flow

(1) **Accurate Prediction of Large-Scale Separated Flow**

Under the combined effects of strong adverse pressure gradients and complex spatial structures with multiple wall surfaces, the internal flow of compressors exhibits characteristics such as multiscale, highly nonlinear coupling, which poses significant challenges for accurately predicting and understanding high-load compressor flows. Although Reynolds-averaged Navier–Stokes (RANS) numerical methods have been widely used in the engineering design of compressors, their use of turbulence models based on isotropic assumptions makes them inadequate for predicting large-scale and highly anisotropic flow structures, such as corner separations [32, 39, 40]. This is evident in their inability to predict transitional behaviors related to separation range and flow losses (as shown in Fig. 1.13), the inability to capture the complex vortex structures' generation and development, and the significant weakening of unsteady effects. These limitations greatly hinder the understanding of the physical essence of phenomena such as corner separation.

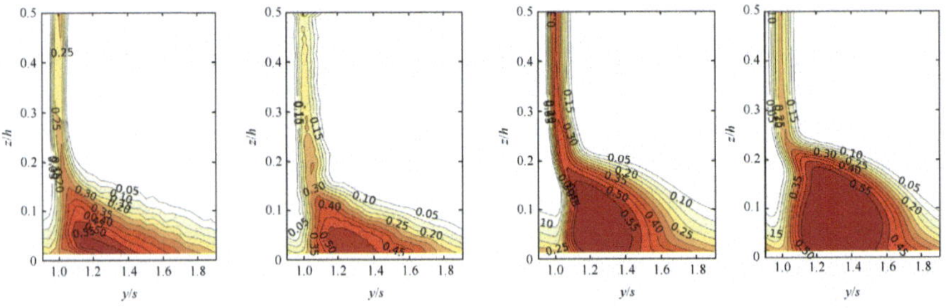

(a) Experiment (b) Large eddy simulation, LES (c) RANS,Wilcox κ-ω (d) RANS, S-A

Fig. 1.13 Comparison of outlet total pressure loss in compressor blade rows [41]

(2) Capturing the Fine Structure of Narrow Multi-surface Flow Fields

In compressors, a clearance gap exists at the blade tip to allow for relative motion between the rotor and the casing. The gap size is typically on the order of millimeters, and the airflow forms leakage vortices and their induced structures are very close to the wall after passing through the narrow space. Additionally, with the application of transonic and supersonic compressors in the twenty-first century, shock waves are inevitably present near the blade tips. When simulating flow fields in geometrically narrow multi-wall surfaces with shock waves, two main challenges arise:

- Compromising between computational cost and accuracy: The results obtained from RANS turbulence models struggle to accurately predict the vortex structures near the wall with strong anisotropy. Besides, high Reynolds number and the multiple wall structures significantly increases the computational demands of LES methods. Furthermore, the ambiguous definition of the interface between RANS and LES limits the ability to resolve vortex structures that are close to the wall.
- Compromising between shockwave capturing accuracy and numerical stability: The presence of shock waves imposes stringent demands on spatial discretization schemes. A second-order accuracy in the discretization of convective terms introduces significant numerical dissipation, causing shock waves to be diffused. On the other hand, high-precision schemes exhibit poor stability and struggle to achieve numerical convergence in high-pressure gradient environments.

In summary, simulating the flow fields in narrow multi-wall surfaces with shock waves presents significant computational challenges and requires high-fidelity methods.

(3) The Contradiction Between Computational Complexity and Accuracy in Multiscale, High-Reynolds Number Flow Fields

The contradiction between computational complexity and accuracy has always been a challenge in numerical simulations. However, for multiscale, high-Reynolds number flow fields inside compressors, this contradiction becomes more pronounced. To meet the demand for fine-scale flow field capturing in the design of next-generation compressors, high-fidelity numerical methods like Large Eddy Simulation (LES) have become a trend. Nevertheless, the internal flow field of compressors spans three to four orders of magnitude in both time and space scales. On the spatial scale, this requires very small grid sizes to capture fine turbulent fluctuations in high-Reynolds number flow fields while also necessitating a large spatial domain to cover the size of the studied object. Similar challenges exist on the temporal scale, where small physical time steps (CFL ~ 1) are needed to resolve high-frequency turbulence, while long simulation times are required to capture large-scale, low-frequency fluctuations. Thus, from both spatial and temporal perspectives, fine-scale numerical simulations of internal compressor flow fields demand significant computational resources, which can be quite substantial given the current state of computing capabilities.

1.3 Numerical Methods for Complex Internal Compressor Flow

1.3.1 Inviscid Numerical Simulation Methods

The numerical simulation of internal compressor flows has undergone a process of development from simple to complex, from approximate to accurate, and from inviscid to viscous. The numerical simulation of complex 3D flow fields inside compressors began after Professor Wu Zhonghua in China proposed the theory of two types of relative flow surfaces (referred to as S1 and S2 flow surfaces) in 1952. This theory is based on an inviscid model and decomposes the 3D flow field into 2D problems on these flow surfaces, greatly simplifying the calculations. Under the extremely limited computational resources available at the time, this approach was well-suited for engineering design and analysis. Consequently, research on compressor design and analysis based on relative flow surface theory experienced significant development [42, 43]. Additionally, inviscid flow models like the Euler equation model and potential function model were gradually applied to numerical simulations of internal fluid flows in fluid machinery [44, 45]. Compared to the previous model, this approach is closer to the three-dimensional characteristics of internal compressor flows. While inviscid flow models do not consider fluid viscosity, they were successful to some extent in simulating internal compressor flows based on the Euler equation method under the computational conditions of that time.

However, inviscid flow models, while changing the physical properties of the fluid, also alter the flow governing equations and their boundary conditions fundamentally. It is challenging to accurately predict phenomena like the shedding of blade trailing edge vortices, boundary layer development, and separation caused by viscosity. Therefore, physical models for internal compressor flow must take viscosity into account to further improve the analysis and design of compressors. The boundary layer/potential flow iterative method [10–12] was one of the earliest numerical simulation methods to introduce viscosity into the fluid governing equations. This method assumes that the main flow region is inviscid potential flow, considering the effects of fluid viscosity only within the boundary layer near the solid wall. The solution is obtained by coupling these two flow regions at the internal boundary. While the boundary layer/potential flow iterative method improved the numerical simulation of compressor internal flow compared to inviscid flow models, it still struggled to predict the complex 3D flow structures caused by viscosity.

1.3.2 Reynolds-Averaged Navier–Stokes (RANS) Numerical Methods

Since the 1980s, with the rapid development of computer capabilities, research on numerical simulations of fully three-dimensional viscous flow inside compressors has made significant progress. Foreign scholars such as Denton [43], Rai [46, 47], Dawes [48], Moore [49], McDonald [50], Hah [51–54], Giles [55], combined with turbulence models, conducted extensive work on studying complex internal flow in compressors by solving the Reynolds-averaged Navier–Stokes (RANS) equations. Domestic scholars [23–35] have also conducted research in this field, providing in-depth analysis and understanding of the flow field structures and characteristics inside compressors. Currently, RANS-based numerical methods have been widely applied in the engineering design of compressors and have played an important role in understanding internal flow fields.

However, numerous studies have shown that predictions based on RANS turbulence models do not align well with experimental results in regions of large separation, such as the corner separation near the blade root, as shown in Fig. 1.13. The differences in predictions of turbulence viscosity and total pressure loss become evident, making it challenging to achieve fidelity in numerical results.

The reason for this issue lies in the Boussinesq assumption used by widely adopted RANS turbulence models, which assumes a linear relationship between the Reynolds stress tensor and the mean strain rate. This assumption is based on fully developed, isotropic, and equilibrium turbulence [56]. The isotropic turbulence assumption restricts the turbulence viscosity from having directional properties, while the equilibrium assumption specifies that the energy production and dissipation within the boundary layer at zero gradient reach an equilibrium state. Additionally, turbulence driven by uniform shear forces exhibits asymptotic equilibrium between energy production and dissipation [57]. However, research has shown that turbulence in regions of large separation exhibits strong anisotropy [58], with increasing anisotropy as the separation strength increases. Moreover, the regions of corner separation, wake, and non-equilibrium phenomena are significant, and these non-equilibrium effects persist downstream of the flow field. Therefore, fundamentally, Boussinesq-based turbulence models are not suitable for predicting complex flows like large separation.

To address this issue, some researchers have made attempts. For example, Monier et al. [59] used a quadratic constitutive relation (QCR) to calculate the Reynolds stress instead of the Boussinesq constitutive relation. The QCR includes the influence of anisotropy when calculating the Reynolds stress using local turbulence viscosity. While the QCR improved the accuracy of predictions, primarily in the upstream flow field and corner separation, it had little effect on correcting the flow field at the exit. Liu Yangwei and others used helicity to modify the S-A model, and the results showed that the modified S-A model effectively improved the accuracy of simulating corner separation in compressors on a time-averaged level. This is beneficial for rapid flow field predictions but is limited by the method itself and cannot resolve fine flow structures in the field.

1.3.3 Development of High-Fidelity Numerical Methods

With the continuous development of high-fidelity numerical methods such as Direct Numerical Simulation (DNS) and Large Eddy Simulation (LES), it has become possible to capture the fine structures of anisotropic and non-equilibrium turbulence in high-load compressor internal flow.

DNS Method [60] involves solving the Navier–Stokes (N-S) equations directly without introducing any turbulence models. All turbulence information, such as Reynolds stress, turbulence kinetic energy, and dissipation rate, is computed directly. The grid scale in DNS needs to be smaller than the smallest eddy scale, known as the dissipation scale, as eddies smaller than this scale dissipate due to viscosity. Therefore, DNS provides turbulence information without any assumptions, and its accuracy can be guaranteed. However, for high Reynolds numbers, the dissipation scale becomes very small. In practical engineering problems of compressors, the Reynolds numbers are typically on the order of 10^6–10^7. Estimated grid sizes for DNS are on the order of 10^{16}. Given the current state of computational resources, these requirements are challenging to meet. As a result, DNS remains difficult to implement and apply in engineering, and existing research is limited to simple geometries and low Reynolds numbers ($Re \sim 10^4$) flows.

LES Method [61] requires significantly less computational effort compared to DNS because it only resolves the large-scale eddy structures in turbulence, while the small-scale eddies in the dissipation region are modeled. The "large scale" here is relative to the small-scale eddy structures in the dissipation region. From an energy cascade perspective, as shown in Fig. 1.14, LES resolves approximately 80% of turbulence kinetic energy, while the remaining 20% is modeled. LES retains anisotropy in the large-scale eddy structures, while the small-scale eddies that are modeled can exhibit isotropic characteristics. Using models to simulate the dissipation of these small-scale eddies can achieve reasonably accurate results. However, when applying LES to solve high-Reynolds-number problems such as compressor internal flow, there is still a challenge related to the huge computational grid requirement, especially in the near-wall boundary layer. Turbulence fluctuations near the walls are very small, and the grid resolution needed to resolve 80% of the turbulence kinetic energy in small-scale eddy structures far exceeds the number of large-scale eddy structures with the same kinetic energy. In 2015, Gao et al. [61] conducted numerical simulations of a low-speed compressor cascade with a Reynolds number of about 10^4 using Large Eddy Simulation (LES). However, LES is still challenging to apply to compressor flow problems confirmed in engineering due to grid requirements.

In recent years, the development of Hybrid RANS/LES methods has provided a new approach for high-fidelity numerical simulations of complex compressor internal flow. The idea behind this approach is to use RANS within the boundary layer and

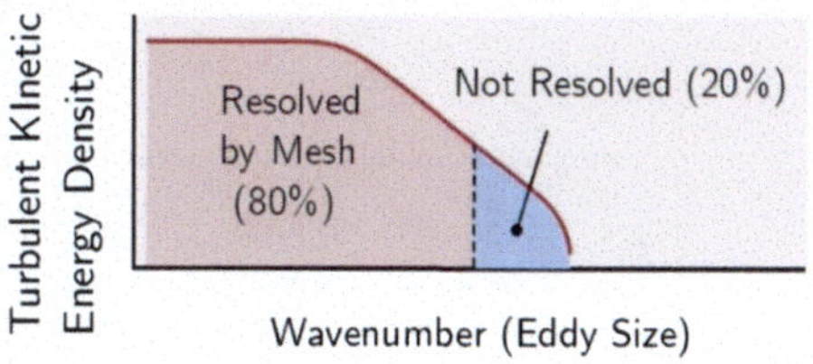

Fig. 1.14 Large eddy simulation method resolving 80% of turbulent kinetic energy [62]

LES in the regions far from the boundary layer, combining the advantages of RANS's computational efficiency within the boundary layer and LES's more accurate prediction of large-scale separation problems away from the wall. This approach holds great promise for engineering applications in the numerical simulation of complex internal flow fields in compressors. Tucker [63] estimated the computational differences between LES and Hybrid RANS/LES methods. As shown in Fig. 1.15, in the numerical simulation of compressors, the Hybrid RANS/LES method can reduce computational requirements by 1–2 orders of magnitude compared to LES alone. This makes it a very promising method for numerically predicting high Reynolds number large separation problems in compressors.

Jochen and Dominic [64] categorized Hybrid RANS/LES methods into two classes based on whether RANS and LES are solved independently. This, along with the second-generation URANS method (traditional RANS methods used for unsteady calculations are generally referred to as the first-generation URANS method), results in a total of three categories, as shown in Fig. 1.16.

The second-generation URANS method maintains the same formal structure as the first-generation URANS but exhibits some LES characteristics in the form of

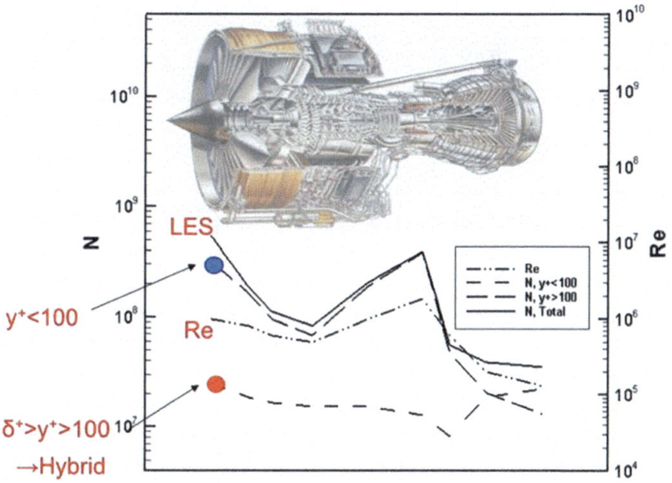

Fig. 1.15 Computational workload for different regions of an engine using LES and RANS/LES coupling methods [63]

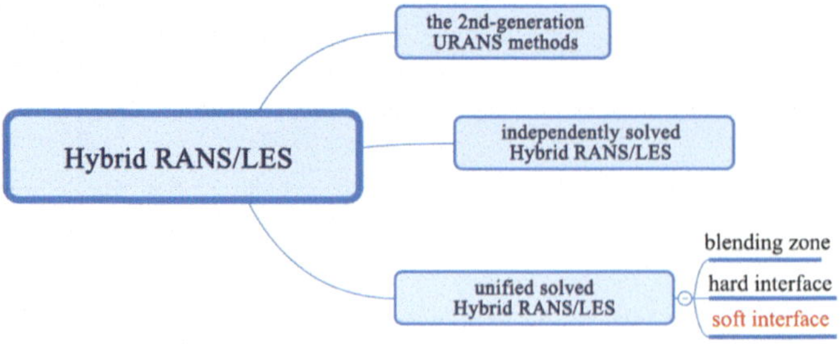

Fig. 1.16 Classification of hybrid RANS/LES methods

Reynolds stress in the mainstream separation region. The most representative method is PANS (Partially Averaged Navier–Stokes), introduced by Girimaji [65]. The idea behind PANS is to modify certain coefficients in the turbulence model based on the RANS method, allowing it to capture finer details of the flow field.

The concept of independently solved Hybrid RANS/LES methods [66] involves using RANS and LES methods separately in different regions of the flow field, with artificial interfaces defined between them. Different turbulence models are used for numerical calculations in each region. This approach has the advantage of avoiding the issues at the interface between RANS and LES in the unified-solution Hybrid RANS/LES methods. However, it has drawbacks such as discontinuous variations in turbulence parameters across the interface and the need for strong empirical guidance, making it unsuitable for complex geometries.

The unified-solution Hybrid RANS/LES methods are constructed based on the similarity between the basic forms of RANS and LES equations. These methods involve using a unified solving approach across the entire domain and employing characteristics of the flow field itself, such as the comparison between grid scale and the closest distance from grid cells to the wall, to transition between turbulence models and subgrid models in different regions. Depending on how the interface between RANS and LES regions is handled, unified-solution Hybrid RANS/LES methods are further subdivided into three types: blending zone methods, hard interface methods, and soft interface methods.

The Detached Eddy Simulation (DES) method, which is studied in this book, is a typical representative of unified-solution Hybrid RANS/LES methods and falls under the soft interface category. It was introduced by Spalart et al. [67] in 1997. DES can automatically partition between RANS and LES and requires minimal changes to the program. As a result, it has gained popularity in computational fluid dynamics and has been gradually applied to the numerical simulation of internal compressor flows. The aim is to achieve high-fidelity simulations of large-scale separated flows near the instability threshold of compressors, potentially enhancing the capture of fine flow structures and supporting research into the underlying mechanisms of compressor flows [39, 41, 68–70].

References

1. LaGraff J E, Ashpis D E, Oldfield M L G, et al. Unsteady flows in turbomachinery, 2006 Workshop on Unsteady Flows in Turbomachinery: NASA, 2006: NASA/CP—2006-214484.
2. Thiam A H, Whittlesey R W, Wark C E and Williams D R. Corner separation and the onset of stall in an axial compressor. Seattle: 38th Fluid Dynamics Conference and Exhibit, 2008.
3. Denton, J D. Loss mechanisms in turbomachines. Journal of Turbomachinery, 1993, 115(4): 621–656.
4. Deng X Y. Numerical investigation on tip clearance flow in compressor. Beijing: Institute of Engineering Thermophysics, Chinese Academy of Sciences, 2006 (in Chinese).
5. Lu X G. Numerical Simulation of Internal Flow in an Axial Flow Compressor Rotor and Analysis of Tip Clearance Leakage. Xi'an: Northwestern Polytechnical University, 2004.
6. Li C Q. Investigation of tip leakage flow in low-speed axial compressor cascade. Beijing: Chinese Academy of Sciences, 2011 (in Chinese).
7. Rains D A. Tip clearance flows in axial flow compressors and pumps. Pasadena: Hydrodynamics and Mechanical Engineering Laboratories, California Institute of Technology, 1954.
8. Lakshminarayana B. Methods of predicting the tip clearance effects in axial flow turbomachinery. Journal of Basic Engineering, 1970, 92(3): 467–480.
9. Chen G T, Greitzer E M, Tan C S, et al. Similarity Analysis of Compressor Tip Clearance Flow Structure. Journal of Turbomachinery, 1991, 113(2): 260–271.
10. Brandt H, Fottner L, Saathoff H, et al. Effects of the inlet flow conditions on the tip clearance flow of an isolated compressor rotor. Amsterdam: Proceedings of ASME Turbo Expo 2002, 2002.
11. Inoue M, Kuroumaru M. Structure of tip clearance flow in an isolated axial compressor rotor. Journal of Turbomachinery, 1989, 111(2): 250–256.
12. Schlechtriem S, and Lotzerich M. Breakdown of tip leakage vortices in compressors at flow conditions close to stall. Orlando: International Gas Turbine & Aeroengine Congress & Exhibition, 1997.
13. Storer J A, Cumpsty N A. Tip leakage flow in axial compressors. Journal of Turbomachinery, 1991, 113(2): 252–259.
14. Ma H W, Jiang H K, Ye D J, et al. Three-dimensional characteristics of turbulent flow in tip region inside an axial-flow compressor rotor passage. Journal of Engineering Thermophysics, 1999, 20(2): 166–170 (in Chinese).
15. Gao L, Li R, Miao F, et al. Unsteady investigation on tip flow field and rotating stall in counter-rotating axial compressor. Journal of Engineering for Gas Turbines and Power, 2015, 137(7): 1–11.
16. Furukawa M, Saiki K, Yamada K, et al. Unsteady flow behavior due to breakdown of tip leakage vortex in an axial compressor rotor at near-stall condition. Munich: Proceedings of the ASME Turbo Expo 2000, 2000.
17. Yamada K, Funazaki K, Furukawa M. The behavior of tip clearance flow at near-stall condition in a transonic axial compressor rotor. Montreal : Proceedings of the ASME Turbo Expo 2007, 2007.
18. Du J. Investigation on the unsteady mechanism of tip leakage flow in transonic compressor/fan rotors. Beijing: Chinese Academy of Sciences, 2010 (in Chinese).
19. Furukawa M, Inoue M, Saiki K, et al. The role of tip leakage vortex breakdown in compressor rotor aerodynamics. Journal of Turbomachinery, 1999, 121(3): 469–480.
20. Garnier V H. Rotating waves as a stall inception indication in axial compressors. Journal of Turbomachinery, 1991, 113(2): 290–301.
21. Day I J. Stall inception in an axial flow compressor. Journal of Turbomachinery, 1993, 115(1): 1–9.
22. Gbadebo, S A, Cumpsty, N A, Hynes, T P. Three-dimensional separation in axial compressors. Journal of Turbomachinery, 2005, 127(2): 331–339.
23. Lei V M, Spakovszky Z S, Greitzer E M. A criterion for axial compressor hub-corner stall. Journal of Turbomachinery, 2008, 130(3): 1–10.

24. Schulz H D, Gallus H D, Lakshminarayana B. Three-dimensional separated flow field in the endwall region of an annular compressor cascade in the presence of rotor-stator interaction: Part I Quasi-steady flow field and comparison with steady-state data. Journal of Turbomachinery, 1990, 112(4): 669–678.
25. Yu X, Liu B. A prediction model for corner separation/stall in axial compressor. Glasgow: Proceedings of ASME Turbo Expo 2010, 2010.
26. Hah C, Loellbach J. Development of hub corner stall and its influence on the performance of axial compressor blade rows. Journal of Turbomachinery, 1999, 121: 67–77.
27. Lewin E, Stark U. Experimental and numerical analysis of hub-corner stall in compressor cascades. Glasgow: Proceedings of the ASME Turbo Expo 2010, 2010.
28. Dong Y, Gallimore SJ and Hodson HP. Three-dimensional flows and loss reduction in axial compressors. Journal of Turbomachinery, 1987, 109: 354–361.
29. Zhang H D, Wu Y, Li Y H, et al. Investigation of vortex structure and flow loss in a high-speed compressor cascade. Acta Aeronautica et Astronautica Sinica, 2014, 35(9): 2438–2450 (in Chinese).
30. Zhao X H, Li Y H, Wu Y, et al. Separation structure and plasma flow control on highly loaded compressor cascade. Acta Aeronautica et Astronautica Sinica, 2012, 33(2): 208–219 (in Chinese).
31. Wang H, Lin D, Su X, et al. Entropy analysis of the interaction between the corner separation and wakes in a compressor cascade. Entropy, 2017, 19(7): 1–20.
32. Xia G, Medic G, Praisner T J. Hybrid RANS/LES simulation of corner stall in a linear compressor cascade. Journal of Turbomachinery, 2018, 140(8): 1–11.
33. Gbadebo S A, Three-Dimensional Separations in Axial Compressors. Cambridge: University of Cambridge, 2004.
34. Gao, F. Advanced numerical simulation of corner separation in a linear compressor cascade. Ecole Centrale de Lyon, Lyons, France. Ecole Centrale de Lyon, 2014.
35. Chen P P, Qiao W Y, Liesner K, et al. Location effect of boundary layer suction on compressor hub-corner separation. Dusseldorf: Proceedings of ASME Turbo Expo 2014, 2014.
36. Liu Y, Yan H, Lu L, et al. Investigation of vortical structures and turbulence characteristics in corner separation in a linear compressor cascade using DDES. Journal of Fluids Engineering, 2017, 139(2): 1–14.
37. Ma W, Ottavy X, Lu L, et al. Intermittent corner separation in a linear compressor cascade. Experiments in fluids, 2013, 54: 1546(6): 1–17.
38. Zambonini G, Ottavy X. Unsteady pressure investigations of corner separated flow in a linear compressor cascade. Montreal: Proceedings of the ASME Turbo Expo 2015, 2015.
39. Liu Y, Zhong L, Lu L. Comparison of DDES and URANS for unsteady tip leakage flow in an axial compressor rotor. Journal of Fluids Engineering, 2019, 141(12): 1–47.
40. Gao Y, Liu Y, Lu L. Investigation of turbulence characteristics in a tip leakage flow model using large-eddy simulation. Phoenix: Proceedings of the ASME Turbo Expo 2019, 2019.
41. Gao F, Ma W, Sun J, et al. Parameter study on numerical simulation of corner separation in LMFA-NACA65 linear compressor cascade. Chinese Journal of Aeronautics, 2017, 30(1): 15–30.
42. Wu CH. A General Theory of Two- and Three-Dimensional Rotational Flow in Subsonic and Transonic Turbomachines. NASA Contractor Report: NASA, 1993: NASA-CR-4496.
43. Denton JD. A time marching method for two- and three-dimensional blade-to-blade flows. Marchwood Engineering Laboratories Report: Aeronautical Research Council, 1975: 3775.
44. Adler D. Status of centrifugal impeller internal aerodynamics - Part 1: inviscid flow prediction methods. Journal of Engineering for Power, 1980, 102(3): 728–737.
45. Adler D, Krimerman Y. On the relevance of inviscid subsonic flow calculations to real centrifugal impellers flow. Journal of Fluids Engineering, 1980, 102(1): 78–84.
46. Rai MM. Navier-Stokes simulations of rotor/stator interaction using patched and overlaid grids. Journal of Propulsion and Power, 1987, 3(5): 387–396.
47. Rai MM. Three-dimensional Navier-Stokes simulations of turbine rotor-stator interaction. Part I-Methodology. Journal of Propulsion and Power, 1989, 5(3): 305–311.

48. Dawes WN. Development of a 3D Navier Stokes Solver for Application to all Types of Turbomachinery. Amsterdam: Proceedings of the ASME 1988 International Gas Turbine and Aeroengine Congress and Exposition, 1988.
49. Moore J, Moore JG, Timmis PH. Performance evaluation of centrifugal compressor impellers using three-dimensional viscous flow calculations. Journal of Engineering for Gas Turbines and Power, 1984, 106(2): 475–481.
50. Briley WR, McDonald H. Solution of the multidimensional compressible Navier-Stokes equations by a generalized implicit method. Journal of Computational Physics, 1977, 24(4): 372–397.
51. Hah C. Calculation of three-dimensional viscous flows in turbomachinery with an implicit relaxation method. AIAA Journal of Propulsion and Power, 1987, 3(5): 415–422.
52. Hah C, Krain H. Secondary flows and vortex motion in a high-efficiency backswept impeller at design and off-design conditions. ASME Journal of Turbomachinery, 1990, 112(1): 7–13.
53. Hah C, Loellbach J. Development of hub corner stall and its influence on the performance of axial compressor blade rows. Journal of Turbomachinery, 1999, 121(1): 67–77.
54. Hah C, Wennerstrom AJ. Three-dimensional flowfields inside a transonic compressor with swept blades. Journal of Turbomachinery, 1991, 113(2): 241–250.
55. Giles MB. Stator/rotor interaction in a transonic turbine. AIAA Journal of Propulsion and Power, 1990, 6(5): 621–627.
56. Wilcox D. Turbulence modeling—an overview. Reno: 39th Aerospace Sciences Meeting and Exhibit, 2001.
57. Fang L, Zhao H K, Lu L P, et al. Quantitative description of non-equilibrium turbulent phenomena in compressors. Aerospace Science and Technology, 2017, 71: 78–89.
58. Yan H, Liu Y, Lu L. Turbulence anisotropy analysis in a highly loaded linear compressor cascade. Aerospace Science and Technology, 2019, 91: 241–254.
59. Monier J F, Poujol N, Laurent M, et al. LES Investigation of Boussinesq constitutive relation validity in a corner separation flow. Oslo: Proceedings of the ASME Turbo Expo 2018, 2018.
60. Zhu H T. Direct Numerical Simulation of Cascade Flow. Beijing: Beihang University, 2009 (in Chinese).
61. Gao F. Advanced numerical simulation of corner separation in a linear compressor cascade. Ecole Centrale de Lyon, 2014.
62. Wimshurst A. Large eddy simulation, Fluid Mechanics 101 [EB/OL], 2021, https://www.fluidmechanics101.com/pages/lectures.html
63. Tucker P G. Computation of unsteady turbomachinery flows: Part 2—LES and hybrids. Progress in Aerospace Sciences, 2011, 47(7): 546–569.
64. Froehlich J, von Terzi D. Hybrid LES/RANS methods for the simulation of turbulent flows. Progress in Aerospace Sciences, 2008, 44(5): 349–377.
65. Girimaji S S. Partially-averaged Navier-Stokes model for turbulence: A Reynolds-averaged Navier-Stokes to direct numerical simulation bridging method. Journal of Applied Mechanics-Transactions of the ASME, 2006, 73(3): 413–421.
66. Quéméré P, Sagaut P. Zonal multi-domain RANS/LES simulations of turbulent flows. International Journal for Numerical Methods in Fluids, 2002, 40(7): 903–925.
67. Spalart P R, Jou W H, Strelets M, et al. Comments on the feasibility of LES for wings and on the hybrid RANS/LES approach. Ruston: Advances in DNS/LES, 1997.
68. Saito S, Yamada K, Furukawa M, et al. Flow structure and unsteady behavior of hub-corner separation in a stator cascade of a multi-stage transonic axial compressor. Oslo: Proceedings of the ASME Turbo Expo 2018, 2018.
69. Yamada K, Furukawa M, Tamura Y, et al. Large-scale detached-eddy simulation analysis of stall inception process in a multistage axial flow compressor. Journal of Turbomachinery, 2017, 139(7): 1–11.
70. Li R, Gao L, Zhao L, et al. Dominating unsteadiness flow structures in corner separation under high Mach number. AIAA Journal, 2019, 57(7): 2923–2932.
71. Li R, Gao L, Ma C, et al. Corner separation dynamics in a high-speed compressor cascade based on detached-eddy simulation. Aerospace Science and Technology, 2020, 99: 1–11.

Chapter 2
Flow Control Equations and DES-Series Methods

2.1 Introduction

The motion of any fluid follows three basic laws, namely, mass conservation, momentum conservation and energy conservation. In the range of Newtonian fluid, all important flow phenomena can be expressed by Navier–Stokes equations (hereinafter referred to as N-S equations). As such, the Navier–Stokes equations are also known as the control equations for flow, and when integrated with the ideal gas law, the equation set becomes complete. However, due to the nonlinearity and the complexity of the geometric structure of fluid regions, the majority of engineering practical flow problems do not have analytical solutions. With the advancement of computer science, the computational fluid dynamics (CFD) has made it possible to solve the N-S equations discretely under complex boundary conditions. In the process of the CFD, the time and space are discretized, and the integral and differential terms in the governing equations are approximated in discrete algebraic form. In the CFD, both time and space are discretized. As a result, the terms in the governing equations are represented in approximate discrete algebraic forms, effectively transforming differential equations into difference equations. By solving these difference equations, discrete solutions for the physical quantities are obtained, which can be used to predict the flow field. This process is referred to as numerical solution. Numerical solution methods in CFD are not unique; however, all methods involve a trade-off between computational effort and accuracy. Therefore, one of the key focuses in the study of computational fluid dynamics is to select the most cost-effective solution method based on the specific physical problem.

Detached Eddy Simulation (DES) combines the advantages of the lower computational cost of Reynolds Averaged Navier–Stokes (RANS) and the precise resolution of large-scale eddies provided by Large Eddy Simulation (LES). For complex flow problems within compressors, it is one of the most promising methods with a high "cost-effectiveness" for practical engineering applications. In this chapter, the

R. Li et al., *Detached-Eddy Simulation Series Methods for Compressor Internal Flow*,
https://doi.org/10.1007/978-981-96-0255-1_2

proposal, idea and development of DES method will be discussed in detail, which will lay a foundation for the application of this method in the field of compressor.

2.2 Basic Governing Equations of Compressible Flow

The integral conservation type governing equation is expressed as:

$$\frac{\partial}{\partial t}\int_{\Omega} \mathbf{U} d\Omega + \int_{S} \mathbf{F}\cdot d\mathbf{S} - \int_{S} \mathbf{F}_V \cdot d\mathbf{S} = \int_{\Omega} \mathbf{S}_T d\Omega \tag{2.1}$$

In the formula:

Ω control volume;
$\mathbf{F}$ inviscid flux vector;
$\mathbf{S}$ surface vector of the control surface.;
$\mathbf{F}_V$ viscous flux vector;
$\mathbf{U}$ conserved variables;
$\mathbf{S}_T$ source item.

Theoretically, Eq. (2.1) can be solved in any coordinate system. This book gives the representation forms in the rectangular coordinate system, the cylindrical coordinate system and any coordinate system based on tensor representation.

2.2.1 *Governing Equations in Cartesian Coordinate System*

The rectangular coordinate system has the advantages of simplicity, intuitiveness and low programming difficulty. In the relative coordinate system, ignoring the body force and the exchange of heat with the outside world, the solution variables, inviscid flux and viscous flux are expressed as (2.2) ~ (2.3) respectively. The tensor representation method is used here. The subscripts "*i*" and "*j*" have a value range (1, 2, 3), which meets the Einstein summation convention and represents the rectangular coordinates respectively. The three directions *x*, *y* and *z* under the system represent the flow direction, tangential direction and span direction respectively when solving the compressor flow field.

$$\mathbf{U} = \begin{bmatrix} \rho \\ \rho w_x \\ \rho w_y \\ \rho w_z \\ E \end{bmatrix} \quad E = \frac{p}{\gamma - 1} + \frac{\rho}{2}\mathbf{W}\cdot\mathbf{W} \tag{2.2}$$

$$\vec{\mathbf{F}} = \begin{bmatrix} \rho\mathbf{W} \\ \rho w_x\mathbf{W} + p\mathbf{e}_x \\ \rho w_y\mathbf{W} + p\mathbf{e}_y \\ \rho w_z\mathbf{W} + p\mathbf{e}_z \\ (E+p)\mathbf{W} \end{bmatrix}, \quad \vec{\mathbf{F}}_V = \begin{bmatrix} 0 \\ \boldsymbol{\tau}\cdot\mathbf{e}_x \\ \boldsymbol{\tau}\cdot\mathbf{e}_y \\ \boldsymbol{\tau}\cdot\mathbf{e}_z \\ \boldsymbol{\tau}\cdot\mathbf{W} + \kappa\nabla T \end{bmatrix}, \tag{2.3}$$

$$\boldsymbol{\tau} = (\mu_l + \mu_t)\left[\nabla\cdot\mathbf{W} + (\nabla\cdot\mathbf{W})^{\mathrm{T}} - \frac{2}{3}(\nabla\cdot\mathbf{W})\cdot\mathbf{I}\right]$$

In the formula:

p pressure;
$\boldsymbol{\tau}$ stress tensor;
T temperature;
w_x, w_y, w_z three components of relative velocity $\mathbf{W}$;
ρ density;
$\mathbf{I}$ identity matrix;
γ Adiabatic index, value 1.4;
κ thermal conductivity, $\kappa = (\mu_l + \mu_t)C_p/Pr$,

Pr is Prandtl number, which is usually 0.708 for ideal air. μ_l and μ_t are laminar and turbulent viscosity coefficients respectively. The Sutherland formula is used to estimate the laminar viscosity coefficient, see (2.4):

$$\mu_l(T) = \mu_\infty\left(\frac{T}{T_\infty}\right)^{1.5}\left(\frac{T_\infty + T_{suth}}{T + T_{suth}}\right) \tag{2.4}$$

Usually $\mu_\infty = 1.716\times10^{-5}$ Pa s, $T_\infty = 293.11$ K, $T_{suth} = 110.55$ K. The turbulent viscosity coefficient μ_t is obtained by solving the turbulence model equation.

Due to the rotation of the coordinate system, the effect of centrifugal force in the momentum equation and energy equation is reflected in the source term, that is:

$$\mathbf{S}_T = \begin{bmatrix} 0 \\ -\rho[2\boldsymbol{\omega}\times\mathbf{W} + (\boldsymbol{\omega}\times(\boldsymbol{\omega}\times\mathbf{r}))] \\ \rho\mathbf{W}\cdot\nabla\left(0.5\omega^2 r^2\right) \end{bmatrix} \tag{2.5}$$

In the formula:

$\boldsymbol{\omega}$ angular velocity vector, its scalar form ω represents the magnitude of angular velocity;
$\mathbf{r}$ circumferential radius vector, x is the rotation axis in this book, in the Cartesian coordinate system $\boldsymbol{\omega}\times\mathbf{r} = \omega y\boldsymbol{e}_z - \omega z\boldsymbol{e}_y$.

In the rectangular coordinate system, it can be expressed as:

$$\mathbf{S}_T = \begin{bmatrix} 0 \\ 0 \\ \omega(2\rho w_z + \omega\rho y) \\ \omega(\omega\rho z - 2\rho w_y) \\ \omega^2(\rho w_y y + \rho w_z z) \end{bmatrix} \tag{2.6}$$

when $\omega = 0$, ignore the body force and the exchange of heat with the outside world, and the source term $\mathbf{S}_T = \mathbf{0}$.

2.2.2 Control Equations in Cylindrical Coordinate System

In the fluid flow problem in the compressor, choosing an appropriate coordinate system is very helpful in solving the problem. According to the characteristics of the impeller rotating around a fixed axis, it is obviously very appropriate to use a cylindrical coordinate system to describe the impeller geometry and internal flow field. The following will give the control equations of the internal flow of the compressor in the relative cylindrical coordinate system.

Usually, the z axis of the cylindrical coordinate system (r, θ, z) is taken to coincide with the rotation axis of the compressor, so that the increasing direction of the θ angle is consistent with the rotation direction of the impeller. Suppose the rotation angular velocity is ω, and the expression of each variable in Eq. (2.1) in the cylindrical coordinate system is Eq. (2.9) ~ (2.11).

$$\mathbf{U} = \begin{bmatrix} \rho \\ \rho w_z \\ r\rho w_\theta \\ \rho w_r \\ E \end{bmatrix} \quad E = \frac{p}{\gamma - 1} + \frac{\rho}{2}\mathbf{W}\cdot\mathbf{W} \tag{2.7}$$

$$\mathbf{F} = \begin{bmatrix} \rho\mathbf{W} \\ \rho w_z\mathbf{W} + p\mathbf{e}_z \\ r\rho w_\theta\mathbf{W} + rp\mathbf{e}_\theta \\ \rho w_r\mathbf{W} + p\mathbf{e}_r \\ (E + p)\mathbf{W} \end{bmatrix}, \quad \mathbf{F}_V = \begin{bmatrix} 0 \\ \boldsymbol{\tau}\cdot\mathbf{e}_z \\ r\boldsymbol{\tau}\cdot\mathbf{e}_\theta \\ \boldsymbol{\tau}\cdot\mathbf{e}_r \\ \boldsymbol{\tau}\cdot\mathbf{W} + \kappa\nabla T \end{bmatrix} \tag{2.8}$$

$$\mathbf{S}_T = \begin{bmatrix} 0 \\ 0 \\ -2\rho\omega r w_r \\ \frac{\rho w_\theta^2}{r} + \rho r\omega^2 + 2\rho\omega w_\theta + \frac{1}{r}(p - \tau_{\theta\theta}) \\ 0 \end{bmatrix} \tag{2.9}$$

Among them, the component expression of the viscous stress tensor $\boldsymbol{\tau}$ in the cylindrical coordinate system is:

$$\begin{aligned}
\tau_{rr} &= -\left(\frac{2}{3}(\mu_l+\mu_t)\left(\frac{\partial w_r}{\partial r}+\frac{\partial w_\theta}{r\partial\theta}+\frac{w_r}{r}+\frac{\partial w_z}{\partial z}\right)\right)+2(\mu_l+\mu_t)\frac{\partial w_r}{\partial r} \\
\tau_{\theta\theta} &= -\left(\frac{2}{3}(\mu_l+\mu_t)\left(\frac{\partial w_r}{\partial r}+\frac{\partial w_\theta}{r\partial\theta}+\frac{w_r}{r}+\frac{\partial w_z}{\partial z}\right)\right)+2(\mu_l+\mu_t)\left(\frac{\partial w_\theta}{r\partial\theta}+\frac{w_r}{r}\right) \\
\tau_{zz} &= -\left(\frac{2}{3}(\mu_l+\mu_t)\left(\frac{\partial w_r}{\partial r}+\frac{\partial w_\theta}{r\partial\theta}+\frac{w_r}{r}+\frac{\partial w_z}{\partial z}\right)\right)+2(\mu_l+\mu_t)\frac{\partial w_z}{\partial z} \\
\tau_{r\theta} &= \tau_{\theta r} = (\mu_l+\mu_t)\left(\frac{\partial w_\theta}{\partial r}+\frac{\partial w_r}{r\partial\theta}-\frac{w_\theta}{r}\right) \\
\tau_{rz} &= \tau_{zr} = (\mu_l+\mu_t)\left(\frac{\partial w_z}{\partial r}+\frac{\partial w_r}{\partial z}\right) \\
\tau_{z\theta} &= \tau_{\theta z} = (\mu_l+\mu_t)\left(\frac{\partial w_\theta}{\partial r}+\frac{\partial w_z}{r\partial\theta}\right)
\end{aligned} \tag{2.10}$$

2.2.3 Control Equations in Any Coordinate System

The general form of the integral conservation type N-S equation of any curvilinear coordinate system is obtained by using tensor analysis operation. The expression is simple and has very strong versatility [1].

Establish any three-dimensional space curve coordinate system $\left(x^1, x^2, x^3\right)$, the corresponding frame vector is (e_1, e_2, e_3), and the inverse component form of the relative velocity vector in this coordinate system is $\mathbf{W} = w^i e_i$. According to the expression form of the high-order tensor field, see Ref. [1] for details, the stress tensor τ_{ij} can be expressed in this coordinate system as:

$$\tau_{ij} = (\mu_l+\mu_t)\left[\left(\nabla^i w^j+\nabla^j w^i\right)-\frac{2}{3}g^{ij}\nabla_i w^i\right]e_i e_j. \tag{2.11}$$

For simplicity in writing, $\sigma^{ij} = (\mu_l+\mu_t)\left[\left(\nabla^i w^j+\nabla^j w^i\right)-\frac{2}{3}g^{ij}\nabla_i w^i\right]$ is added, then $\tau_{ij} = \sigma^{ij} e_i e_j$.

Therefore, in this arbitrary coordinate system, the solution variables, inviscid and viscous flux vectors and source terms in Eq. (2.1) are expressed in the form of tensors:

$$\mathbf{U} = \begin{cases} \rho \\ \rho w^i \\ E \end{cases}, \quad \mathbf{F} = \begin{cases} \rho w^i \\ \rho w^i w^j + p g^{ij} \\ (\rho E + p) w^j \end{cases}, \quad \mathbf{F}_{\mathrm{V}} = \begin{cases} 0 \\ \sigma^{ij} \\ g_{ik} w^k \sigma^{ij} + \kappa g^{ij} \nabla_i T \end{cases},$$
$$\mathbf{S}_{\mathrm{T}} = \begin{cases} 0 \\ \left(\rho w^k w^j - \sigma^{kj} + p g^{kj}\right) \Gamma^i_{kj} + \rho g^{ki} \varepsilon_{mjk} w^m a^j \\ 0 \end{cases}, \tag{2.12}$$

where $a^j = \boldsymbol{\omega} \times \mathbf{r} + 2\mathbf{W} = g^{kj} \varepsilon_{mik} w^m r^i + 2w^j$. In the above tensor relations, represents the contravariant components of the fundamental metric tensor describing spatial geometric properties, g is the determinant of the metric tensor; Γ^i_{jk} denote the Christoffel symbols of the second kind,[1] ε_{ijk} represent Levi–Civita symbols.[2]

The tensor representation can combine the rectangular coordinate system and the cylindrical coordinate system. The values of g^{ij} and Γ^i_{jk} are different in different coordinate systems:

(1) Cartesian Coordinate System

$$g^{11} = 1 \quad g^{22} = 1 \quad g^{33} = 1 \quad g^{ij} = 0\,(i \neq j)$$

$$g_{11} = 1 \quad g_{22} = 1 \quad g_{33} = 1 \quad g_{ij} = 0\,(i \neq j)$$

$$\Gamma^1_{11} = 1,\ \Gamma^2_{22} = 1,\ \Gamma^3_{33} = 1$$

(2) Cylindrical Coordinate System Quantities

$$g^{11} = 1 \quad g^{22} = 1 \quad g^{33} = \frac{1}{r^2} \quad g^{ij} = 0(i \neq j)$$

$$g_{11} = 1 \quad g_{22} = 1 \quad g_{33} = r^2 \quad g_{ij} = 0\,(i \neq j),\, g = r^2$$

$$\Gamma^2_{33} = -r,\ \Gamma^3_{23} = \Gamma^3_{32} = \frac{1}{r}$$

[1] Also called contact coefficient, defined as: $\frac{\partial e_j}{\partial x^j} = \Gamma^i_{jk} e_i$.

[2] Levi–Civita symbol, defined as: $\varepsilon_{ijk} = \begin{cases} 1 & \text{if } (i, j, k) \text{ is an even permutation of } (1, 2, 3) \\ -1 & \text{if } (i, j, k) \text{ is an odd permutation of } (1, 2, 3) \\ 0 & \text{if } i = j = k \end{cases}$.

Table 2.1 Selection of feature parameters

Characteristic parameters	Feature parameter selection	Characteristic parameters	Feature parameter selection
T_{ref}	Inlet total temperature T_{in}^*	V_{ref}	$V_{ref} = \sqrt{\gamma RT_{in}^*}$
ρ_{ref}	$\rho_{ref} = p_{in}^*/(RT_{in}^*)$	p_{ref}	$p_{ref} = \gamma p_{in}^*$
μ_{ref}	Inlet laminar flow viscosity coefficient	L_{ref}	$L_{ref} = 1$, no dimensionless processing

2.2.4 *Nondimensionalization of the Governing Equations*

The physical parameters in the control equations are very different in magnitude, which will cause large numerical errors and loss of accuracy during the numerical solution operation [2]. The control equations need to be dimensionless to improve the fidelity of the numerical results. Spend. Considering that the program developed in this article is mainly used in the numerical simulation of the internal flow of the compressor, based on the characteristic length of the compressor and the physical quantities total temperature T_{in}^* and total pressure p_{in}^* that can be obtained from experimental measurements, the control The physical quantities used in the dimensionless treatment of equations are given in detail in Table 2.1. The subscripts "ref" represents the characteristic parameters for nondimensionalization. It should be noted that the total temperature and total pressure here are values in the absolute coordinate system.

According to the above characteristic parameters, the physical quantities in the governing equations can be dimensionless. The letter superscript "~" represents the dimensionless physical quantities. The dimensionless density, temperature, pressure, speed and viscosity coefficient are respectively[3]:

$$\widetilde{\rho} = \rho/\rho_{ref};\ \widetilde{T} = T/T_{ref};\ \widetilde{p} = p/p_{ref};\ \widetilde{v}_i = v_i/V_{ref};\ \widetilde{w}_i = w_i/V_{ref};\ \widetilde{\mu} = \mu/\mu_{ref} \tag{2.13}$$

The dimensionless rotation angular velocity, time and gas constant are respectively:

$$\widetilde{\omega} = \frac{\omega}{V_{ref}/L_{ref}};\ \widetilde{t} = \left(\frac{t}{L_{ref}/V_{ref}}\right);\ \widetilde{R} = \frac{1}{\gamma} \tag{2.14}$$

Bring the above dimensionless quantities into Eq. (2.1) and merge them to obtain a system of dimensionless equation, see (2.15):

$$\frac{\partial}{\partial \widetilde{t}} \int\limits_{\Omega} \widetilde{\mathbf{U}} d\widetilde{\Omega} + \int\limits_{S} \widetilde{\mathbf{F}} \cdot d\widetilde{\mathbf{S}} - \frac{1}{Re_{ref}} \int\limits_{S} \widetilde{\mathbf{F}}_V \cdot d\widetilde{\mathbf{S}} = \int\limits_{\Omega} \widetilde{\mathbf{S}}_T d\widetilde{\Omega} \tag{2.15}$$

[3] The speed subscript "i" takes (1, 2, 3) and represents the speed component.

In the formula:

Re_{ref} characteristic Reynolds number, $Re_{\mathrm{ref}} = \rho_{\mathrm{ref}} V_{\mathrm{ref}} L_{\mathrm{ref}} / \mu_{\mathrm{ref}}$.

After the control equation variables adopt the dimensionless form, their form does not change, except that the characteristic Reynolds number term is added to the viscous flux term. Removing the superscript "~" will not cause ambiguity. Therefore, in order to simplify the expression, the superscript will be removed in the following introduction of numerical methods as long as it does not cause ambiguity.

2.3 High-Precision Time and Space Discretization Schemes

For different physical problems, it is of great significance to choose reasonable time and space discrete formats for improving fidelity, reducing computational consumption and ensuring computational stability. Based on the special requirements of high-fidelity numerical methods like DES method for spatiotemporal discretization, this section introduces time and space discretization formats suitable for high-subsonic, strong adverse pressure gradient, and unsteady flow fields.

2.3.1 *Time Discretization*

To enhance the temporal discretization accuracy of unsteady flow fields and ensure high computational efficiency, Jameson proposed the dual-time stepping method in 1991 [3]. The fundamental idea is to transform the problem to be solved at each time point into a steady problem for iterative solution. The time advancement includes physical time iteration (outer iteration) and pseudo-time (inner iteration). After adding a fictitious time term to the original governing Eq. (2.1):

$$\int_{\Omega} \frac{\partial \mathbf{U}}{\partial \varsigma} \mathrm{d}\Omega = \int_{\Omega} \mathbf{S}_T \mathrm{d}\Omega - \left(\int_{S} \mathbf{F} \cdot \mathrm{d}\mathbf{S} - \int_{S} \mathbf{F}_V \cdot \mathrm{d}\mathbf{S} \right) - \frac{\partial}{\partial t} \int_{\Omega} \mathbf{U} \mathrm{d}\Omega \tag{2.16}$$

where the ς and t represent pseudo-time and physical time, respectively. Thus, the unsteady problem is transformed into a steady problem. When $\partial \mathbf{U}/\partial \varsigma \rightarrow 0$, it implies that Eq. (2.16) becomes equivalent to the original governing Eq. (2.1). Various acceleration techniques (such as local time stepping and multigrid methods) can be employed to increase the convergence rate, and the increase in the number of inner iterations effectively improves the temporal accuracy.

(1) The Physical Time Discretization

The discretization of the physical time term employs a three-point implicit backward difference scheme, as given in (2.17):

$$\left(\frac{\partial}{\partial t}\int\limits_{\Omega} \mathbf{U}d\Omega\right)^{n+1} = \frac{3\mathbf{U}^{n+1}\Omega^{n+1} - 4\mathbf{U}^{n}\Omega^{n} + \mathbf{U}^{n-1}\Omega^{n-1}}{2\Delta t} \tag{2.17}$$

where n and Δt represents current physical time and the physical time step.

During the iterative process, when the computational grid, does not change with time, i.e., $\Omega^{n+1} = \Omega^{n} = \Omega^{n-1}$. As a result, the physical time step size can be chosen based on the actual unsteady fluctuating flow field, no longer limited by the constraints of computational stability.

(2) The Pseudo Time Discretization

The fourth-order Runge–Kutta display format (hereinafter referred to as R-K) is used for the discretization of the pseudo time term, which has the advantages of high precision and robustness, and has good inheritance in the program, such as (2.18).

$$\begin{aligned} &\mathbf{U}^{(0)} = \mathbf{U}^{k} \\ &\mathbf{U}^{(m)} = \mathbf{U}^{(0)} + CFL \cdot \alpha_m \frac{\Delta\varsigma}{\Omega}\left(\mathbf{R}^{(m-1)} - \frac{\Omega}{2\Delta t}\left(3\mathbf{U}^{(m-1)} - 4\mathbf{U}^{n} + \mathbf{U}^{n-1}\right)\right) \\ &\mathbf{U}^{k+1} = \mathbf{U}^{(4)} \end{aligned} \tag{2.18}$$

where

k current pseudo time step;
$\Delta\varsigma$ pseudo time step size;
m $m \in [1, 2, 3, 4]$;
α_m constant coefficient $\alpha_m = m/4$;
$\mathbf{R}$ residual, $\mathbf{R} = \int_{\Omega} \mathbf{S}_T d\Omega - \left(\int_S \mathbf{F}_I \cdot d\mathbf{S} - \int_S \mathbf{F}_V \cdot d\mathbf{S}\right)$.

It can be seen from the Fourier analysis that the limit courant number of the fourth-order scheme is $2\sqrt{2}$ if the viscosity term is calculated at each time step while 2.6 as only at the first time step. In the calculation of high-speed and strong adverse pressure gradient flow field, the CFL number needs to be further reduced due to the stability of the calculation.

In order to reduce the cost of solving the dissipation term, the calculation of artificial viscosity and physical viscosity is only performed in the first step of time advancement, and frozen in the remaining steps. The pseudo time step has no practical physical meaning. In order to accelerate the convergence, improve the computational efficiency and reduce the number of iterations of the pseudo time layer, the local time step method, implicit residual smoothing and other accelerated convergence techniques can be used.

2.3.2 *Spatial Discretization*

Spatial discretization includes the discretization of inviscid and viscous flux. The viscous term is also called the diffusion term, where the flow field information diffuses from the center to the periphery. The central difference scheme is generally used for viscous term discretization, and the second-order central difference scheme is the most common. The inviscid term, also known as the convection term, plays a decisive role in the numerical accuracy, computational stability and accuracy of the flow field. To some extent, it can be considered that the computational format is the core of numerical simulation, so this section will focus on the spatial discrete format of the convection term.

In the finite volume method, the purpose of the spatial discretization scheme of the convection term is to obtain the flux at the grid surface $J + 1/2$ in Fig. 2.1.

Central Difference Scheme and Artificial Viscosity

The simplest way to solve the flux is based on the average of the parameters on the center point of the adjacent unit, that is, the central difference scheme with second-order accuracy is used to discretize the convection term in the control equation. The second-order central difference scheme has a simple flux calculation logic and is easy to implement in programming. It can be conveniently extended to multi-block grid systems and has been widely used in subsonic and transonic flows [4–6]. It has also become the preferred scheme in many commercial software applications. In the smooth flow field, the computational structure of the second-order central difference scheme has second-order accuracy. However, when the flow field is discontinuous, the accuracy is difficult to be guaranteed.

There are two methods to calculate the flux on the grid surface:

(1) The flux of the center point is obtained according to the conserved quantity of the unit center point, and then the flux of the two center points is averaged to obtain the flux of the interface, that is:

$$\mathbf{F}_{I,J+1/2,K} = \left(\mathbf{F}\left(\mathbf{U}_{I,J,K}\right) + \mathbf{F}\left(\mathbf{U}_{I,J+1,K}\right)\right)/2 \tag{2.19}$$

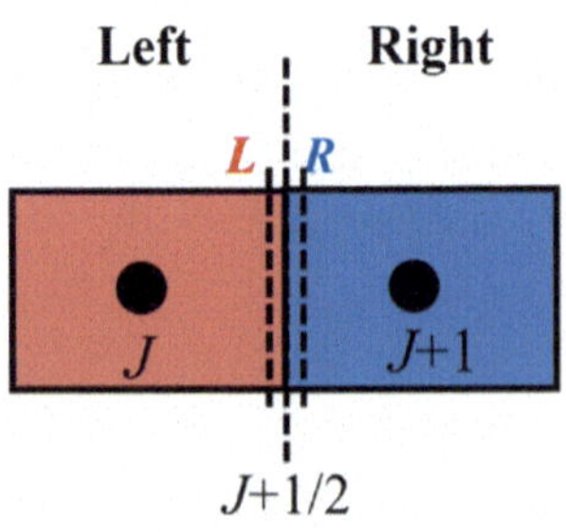

Fig. 2.1 The diagram of flux solution interface

(2) The conserved quantity on the unit surface is obtained by averaging the conserved quantities of the two units center points, and then the flux on the unit surface is calculated, that is

$$\mathbf{F}_{I,J+1/2,K} = \mathbf{F}\big(\big(\mathbf{U}_{I,J,K} + \mathbf{U}_{I,J+1,K}\big)/2\big) \tag{2.20}$$

The formula (2.19) is less computationally intensive, but experience shows that the formula (2.20) is more robust, especially for high-speed flow conditions.

When the central difference scheme is used to discretize the solution domain, it is not dissipative, and some high-frequency error components are not attenuated in the solution process. Therefore, it is necessary to correct or limit the flux on each surface. This correction is usually called artificial dissipative flux to suppress the odd–even disconnection that may be caused by the central scheme and effectively capture the shock wave, to avoid the oscillation phenomenon and make the numerical solution converge to the steady state, so as to obtain a smooth pressure distribution.

This section introduces the adaptive artificial viscosity term [4–6] constructed by Jameson. The basic dissipation term coefficient is a second-order difference term proportional to the absolute value of the second-order derivative of the pressure. This term can capture the shock waves and stagnation points flow and suppress the numerical oscillation near them. However, when the calculation converges to a steady state, some low-frequency small fluctuations cannot be eliminated. Therefore, a fourth-order difference term is introduced as the background numerical viscosity to suppress the odd–even oscillation and increase the dissipation in the smooth region of the flow.

After introducing artificial viscosity, the flux is expressed as:

$$\mathbf{F}_{I,J+1/2,K} = \mathbf{F}\big(\big(\mathbf{U}_{I,J,K} + \mathbf{U}_{I,J+1,K}\big)/2\big) - \mathbf{D}_{I,J+1/2,K} \tag{2.21}$$

In the formula, $\mathbf{D}_{I,J+1/2,K}$ is the corresponding dissipation, and the expression is:

$$\begin{aligned}\mathbf{D}_{I,J+1/2,K} = {} & \varepsilon^{(2)}_{I,J+1/2,K} R_{I,J+1/2,K} \boldsymbol{\Delta}_{I,J+1/2,K} \\ & - \varepsilon^{(4)}_{I,J+1/2,K} R_{I,J+1/2,K} \big(\boldsymbol{\Delta}_{I,J+3/2,K} - 2\boldsymbol{\Delta}_{I,J+1/2,K} + \boldsymbol{\Delta}_{I,J-1/2,K}\big),\end{aligned} \tag{2.22}$$

where $\boldsymbol{\Delta}$ is the difference between the conserved variables on both sides of the interface.

$$\boldsymbol{\Delta}_{I,J+1/2,K} = \mathbf{U}_{I,J+1,K} - \mathbf{U}_{I,J,K}, \tag{2.23}$$

$R_{I,J+1/2,K}$ approximates the spectral radius of the Jacobian matrix of the local flux, which can be taken as

$$R_{I,J+1/2,K} = \big(\Omega_{I,J+1,K}/\Delta t^*_{I,J+1,k} + \Omega_{I,J,K}/\Delta t^*_{I,J,k}\big)/2, \tag{2.24}$$

Δt^* is the local time step when the CFL number is 1. We define:

$$\theta_{I,J,K}^{(J)} = \left| \frac{p_{I,J+1,K} - 2p_{I,J,K} + p_{I,J-1,K}}{p_{I,J+1,K} + 2p_{I,J,K} + p_{I,J-1,K}} \right|, \tag{2.25}$$

$\varepsilon^{(2)}$ and $\varepsilon^{(4)}$ are defined as:

$$\varepsilon_{I,J+1/2,K}^{(2)} = k^{(2)} \max\left\{\theta_{I,J+1,K}^{(J)}, \theta_{I,J,K}^{(J)}\right\}, \quad \varepsilon_{I,J+1/2,K}^{(4)} = \max\left\{0, \left(k^{(4)} - \varepsilon_{I,J+1/2,K}^{(2)}\right)\right\}. \tag{2.26}$$

Here, $k^{(2)}$ and $k^{(4)}$ are the input constants, usually $k^{(2)} = 0.125 \sim 0.5$ while $k^{(4)} = 1/128 \sim 0.025$. When there is no shock wave in the flow field, $k^{(2)} = 0$, the dissipation flux is a third-order small quantity in the smooth region of the solution. While the accuracy of the solution can be guaranteed, the fluctuation will be suppressed, and the convergence can be accelerated. On the contrary, Near the shock wave, the fourth-order smooth term is closed while the second-order dissipative term is on the order of 1. Currently, the scheme degenerates into first-order accuracy, thus effectively suppressing the fluctuations near the shock wave.

Upwind Schemes

The flow field information of the convection term has a clear direction, that is, the flow field information (wave) is transmitted from the upstream to the downstream in that the cell face value is made dependent on the upwind nodal value. Therefore, the upwind format discretization can ensure the stability of the calculation. The key to the application of the upwind format is to find the direction of the 'wind', that is, the direction of information (wave) propagation. For the one-dimensional convection equation ($\partial u/\partial t + a\partial u/\partial x = 0$), because it is a single-wave equation, the direction can be judged by the positive and negative of the coefficient a. However, the N-S equations involve multiple waves, thus necessitating the use of flux splitting techniques to decouple the solution and identify independently propagating waves. The flux difference splitting method is derived based on the characteristic flux differential splitting while satisfying the conservation law of flux, and it has the characteristic of local decoupling. This method has low dissipation, low numerical oscillation, and good stability, making it an excellent choice for compressor problems with strong adverse pressure gradients. Among the flux difference splitting methods, the Roe scheme is the most typical.

1. The Construction Principle of the Roe Scheme

The Roe scheme is known for its low dissipation, high discontinuity resolution, and computational stability, which allows it to achieve both high shock and vortex resolution. These attributes make it highly suitable for high-fidelity predictions of high-subsonic flow fields with large separation, such as those found in transonic flows. Therefore, this book will focus on the Roe scheme within the context of flux difference splitting methods [7]. The core of the Roe schemes is to discretize the derivatives of the solution variables in the flux terms by approximating the solution to the Riemann problem at the interfaces [8], thereby obtaining the fluxes that cross the interfaces.[4] Since this section only considers the discretization method for the convective terms, unrelated to the viscous and source terms, for ease of explanation, let's first consider the one-dimensional Euler equations:

$$\frac{\partial \mathbf{U}}{\partial t} + \frac{\partial \mathbf{F}}{\partial x} = \mathbf{0} \tag{2.27}$$

where

$$\mathbf{U} = \begin{bmatrix} \rho \\ \rho u \\ E \end{bmatrix} \quad \mathbf{F} = \begin{bmatrix} \rho u \\ \rho u^2 + p \\ u(E+p) \end{bmatrix}$$

u represents the one-dimensional velocity. The flux vector $\mathbf{F}$ is a function of the solution variable $\mathbf{U}$, therefore Eq. (2.27) can be written as:

$$\frac{\partial \mathbf{U}}{\partial t} + \mathbf{A} \cdot \frac{\partial \mathbf{U}}{\partial x} = \mathbf{0} \tag{2.28}$$

In which $\mathbf{A} = \partial \mathbf{F}/\partial \mathbf{U}$ is the Jacobian matrix of the system of equations, thus making the form of Eq. (2.28) identical to that of a one-dimensional convection equation. The purpose of the Roe scheme is to find an approximate matrix $\widehat{\mathbf{A}}(\mathbf{U}_L, \mathbf{U}_R)$ that meets the following three conditions:

(1) $\mathbf{F}(\mathbf{U}_R) - \mathbf{F}(\mathbf{U}_L) = \widehat{\mathbf{A}}(\mathbf{U}_R - \mathbf{U}_L)$, i.e., $\widehat{\mathbf{A}}(\mathbf{U}_L, \mathbf{U}_R)$ is the average increase rate of $\mathbf{F}(\mathbf{U})$;

(2) $\widehat{\mathbf{A}}(\mathbf{U}_L, \mathbf{U}_R)$ is continuous and consistent with the Jacobian matrix $\mathbf{A}(\mathbf{U})$, that is $\widehat{\mathbf{A}}(\mathbf{U}, \mathbf{U}) = \mathbf{A}(\mathbf{U})$;

(3) The matrix $\widehat{\mathbf{A}}(\mathbf{U}_L, \mathbf{U}_R)$ possesses eigenvalues and is diagonalizable by means of a similarity transformation.

Within this context, $\mathbf{U}_L$ and $\mathbf{U}_R$ represent the reconstructed variables adjacent to the interface $(J + 1/2)$, as depicted in Fig. 2.1. An elaborate discussion on the reconstruction methodology will be presented in the subsequent section. For the current discussion, the most straightforward reconstruction approach is employed, where $\mathbf{U}_L = \mathbf{U}_{J-1}$ and $\mathbf{U}_R = \mathbf{U}_{J+1}$. Roe introduced an approximation for the matrix, founded upon the Roe average, as referenced in Eq. (2.29). The overbar notation ("–") signifies the application of the Roe averaging technique.

[4] Riemann problem, a specific type of initial value problem that consists of conservation equations and piecewise constant initial conditions. The initial constants are discontinuous at the target interface.

After derivation, the expressions for each term in Eq. (2.32) under the local coordinate system for the three-dimensional problem are as follows:

$$\widehat{\mathbf{F}}_{J+1/2} = \frac{1}{2}(\mathbf{F}_{J+1/2||L} + \mathbf{F}_{J+1/2||R}) - \frac{1}{2}\sum_{k=1}^{5}\alpha_k\left|\overline{\lambda}_k\right|\overline{\varphi}^{(k)}$$

$$\widehat{\mathbf{F}}_{J+1/2||L} = \begin{bmatrix} \rho_L u_{n||L} \\ \rho_L u_{n||L}u_{n||L} + p_L \\ \rho_L u_{n||L}u_{c_1||L} \\ \rho_L u_{n||L}u_{c_2||L} \\ (E_L + p_L)u_{n||L} \end{bmatrix} \quad \widehat{\mathbf{F}}_{J+1/2,R} = \begin{bmatrix} \rho_R u_{n||R} \\ \rho_R u_{n||R}u_{n||R} + p_R \\ \rho_R u_{n||R}u_{c_1||R} \\ \rho_R u_{n||R}u_{c_2||R} \\ (E_R + p_R)u_{n||R} \end{bmatrix}$$

$$\alpha_1 = \frac{1}{2\overline{a}^2}(\Delta p - \overline{\rho a}\cdot\Delta u_n); \quad \alpha_2 = \Delta\rho - \Delta p/\overline{a}^2$$

$$\alpha_3 = \overline{\rho}\cdot\Delta u_{c_1}; \quad \alpha_4 = \overline{\rho}\cdot\Delta u_{c_2}; \quad \alpha_5 = \frac{1}{2\overline{a}^2}(\Delta p + \overline{\rho a}\cdot\Delta u_n)$$

$$\overline{\varphi}^{(1)} = \begin{bmatrix} 1 \\ \overline{u}_n - \overline{a} \\ \overline{u}_{c_1} \\ \overline{u}_{c_2} \\ \overline{H} - \overline{u}_n\overline{a} \end{bmatrix}, \quad \overline{\varphi}^{(2)} = \begin{bmatrix} 1 \\ \overline{u}_n \\ \overline{u}_{c_1} \\ \overline{u}_{c_2} \\ \frac{1}{2}(\overline{u}_n^2 + \overline{u}_{c_1}^2 + \overline{u}_{c_2}^2) \end{bmatrix}, \overline{\varphi}^{(3)} = \begin{bmatrix} 0 \\ 0 \\ 1 \\ 0 \\ \overline{u}_{c_1} \end{bmatrix} \tag{2.35}$$

$$\overline{\varphi}^{(4)} = \begin{bmatrix} 0 \\ 0 \\ 1 \\ 0 \\ \overline{u}_{c_2} \end{bmatrix}, \quad \overline{\varphi}^{(5)} = \begin{bmatrix} 1 \\ \overline{u}_n + \overline{a} \\ \overline{u}_{c_1} \\ \overline{u}_{c_2} \\ \overline{H} + \overline{u}_n\overline{a} \end{bmatrix}$$

$$\overline{\lambda}_1 = \overline{u}_n - \overline{a} \quad \overline{\lambda}_2 = \overline{\lambda}_3 = \overline{\lambda}_4 = \overline{u}_n \quad \overline{\lambda}_5 = \overline{u}_n + \overline{a}$$

where

a sound speed;
u_n, u_{c_1}, u_{c_2} normal and two tangential velocities of the interface.

3. LDE Schemes

One of the challenges associated with the Roe scheme is its substantial computational demand, which necessitates a considerable expenditure of CPU time. In response to this, Zha et al. [9] have advanced the LDE (Low Diffusion E-CUSP) scheme. The underlying principle of this scheme is the decomposition of the inviscid flux into two components: a pressure flux and a conservative flux. This approach effectively precludes the need for intricate matrix operations, thereby conserving a significant portion of computational time. Relative to the Roe scheme, the LDE scheme is characterized by its reduced computational complexity, accelerated convergence rate, and enhanced stability.

The inviscid flux $\boldsymbol{F}$ in the one-dimensional Euler Eq. (2.20) can be decomposed as follows:

$$\mathbf{F} = \mathbf{F}^c + \mathbf{F}^p = \begin{pmatrix} \rho u \\ \rho u^2 \\ Eu \end{pmatrix} + \begin{pmatrix} 0 \\ p \\ pu \end{pmatrix} \tag{2.36}$$

The eigenvalues of the Jacobian matrix of $\mathbf{F}^c$ are (u, u, u), representing the convective terms, and can be rewritten as:

$$\mathbf{F}^c = \rho u \begin{pmatrix} 1 \\ u \\ E/\rho \end{pmatrix} = \rho u \mathbf{q} \tag{2.37}$$

The Jacobian matrix of $\mathbf{F}^p$ is $(-a, 0, a)$, where $a = (\gamma RT)^{\frac{1}{2}}$ denotes the speed of sound, with positive and negative signs indicating the direction of wave propagation. In response to the two components of the aforementioned decomposition, Zha et al. [9] suggest treating $\mathbf{F}^c$ in an upwind manner. The flux $\mathbf{F}^c$ at the interface $J + 1/2$ (for simplicity and to avoid ambiguity, $J + 1/2$ is abbreviated as 1/2) is given by:

$$\mathbf{F}^a_{1/2} = a_{1/2}\left[\rho_L a^+ \mathbf{q}_L + \rho_R a^- \mathbf{q}_R\right] \tag{2.38}$$

where

$$\begin{aligned}
& a_{1/2} = \frac{1}{2}(a_L + a_R) \\
& a^+ = \alpha_L^+(1+\beta_L)M_L - \beta_L M_L^+ - M_{1/2}^+ \quad a^- = \alpha_R^-(1+\beta_R)M_R - \beta_R M_R^- + M_{1/2}^- \\
& \alpha_{L,R}^{\pm} = \frac{1}{2}\left[1 \pm \operatorname{sign}(M_{L,R})\right] \quad \beta_{L,R} = -\max\left[0, 1 - \operatorname{int}(|M_{L,R}|)\right] \\
& M_{L,R} = \frac{U_{L,R}}{a_{1/2}} \quad M_{1/2}^+ = M_{1/2}\frac{a_R + a_L\phi}{a_R + a_L} \quad M_{1/2}^- = M_{1/2}\frac{a_L + a_R\phi^{-1}}{a_L + a_R} \\
& \phi = \frac{\rho_R a_R^2}{\rho_L a_L^2} \quad M_{1/2} = \beta_L \delta^+ M_L^- - \beta_R \delta^- M_R^+ \\
& M_{L,R}^{\pm} = \pm\frac{1}{4}(M_{L,R} \pm 1)^2 \quad \delta^{\pm} = \frac{1}{2}\left\{1 \pm \operatorname{sign}\left[\frac{1}{2}(M_L + M_R)\right]\right\}
\end{aligned} \tag{2.39}$$

And for the propagation direction, $\mathbf{F}^p$ is split into an upwind term and a downwind term, which are then individually weighted and averaged using $u - a$ and $u + a$ as the respective weights. The expression is given by:

$$\mathbf{F}^p = \begin{bmatrix} 0 \\ p \\ pu \end{bmatrix} = \begin{bmatrix} 0 \\ \lambda_L^+ p_L + \lambda_R^- p_R \\ a_{1/2}\left(S_L^+ p_L + S_R^+ p_R\right) \end{bmatrix} \tag{2.40}$$

where

$$\begin{aligned} \lambda_{L,R}^{\pm} &= \left[\alpha(1+\beta) - \beta\chi^{\pm}\right]_{L,R} \\ \chi_{L,R}^{\pm} &= \frac{1}{4}\left(M_{L,R} \pm 1\right)^2\left(2 \mp M_{L,R}\right) \\ S_{L,R}^{\pm} &= \left[\alpha^{\pm}(1+\beta)M_{L,R} - \beta M_{L,R}\right] \\ M_{L,R} &= \frac{u_{L,R}}{a_{1/2}} \end{aligned} \tag{2.41}$$

The conceptual approach for applying the LDE scheme in three-dimensional meshes is analogous to that of the Roe scheme, and thus will not be further elaborated upon here.

Variable Reconstruction

When solving equations using flux-difference splitting discretization schemes such as Roe and LDE, to achieve high precision in the computation of the inviscid fluxes, it is necessary to reconstruct the original variables on the left and right sides of the grid cells using the values at the adjacent cells (i.e., the values at the left and right interfaces of the cells denoted by superscripts L and R in Eqs. (2.29) to (2.35) and (2.38) to (2.41)). These reconstructions are typically obtained using flux limiters, MUSCL interpolation, or WENO reconstruction. The following section will introduce various methods for reconstructing the variables on the left and right sides of the interfaces with different levels of accuracy.

MUSCL interpolation is the most commonly used method for upwind schemes, as it can enhance the accuracy of upwind schemes while maintaining their monotonicity. Flux limiters can restrict high-order terms and effectively prevent numerical oscillations. Flux limiters are typically used in conjunction with MUSCL interpolation and can also be used independently; however, when used independently, they only provide first-order accuracy. If neither MUSCL interpolation nor flux limiters are employed, zeroth-order reconstruction can achieve first-order accuracy. So-called zeroth-order reconstruction involves using a first-order scheme to give the interfaces an upwind characteristic. This section has conducted research on five parts: zeroth-order reconstruction, MUSCL interpolation, MUSCL interpolation combined with flux limiters, the use of flux limiters alone, and fifth-order WENO reconstruction. The forthcoming discussion will illustrate the concepts using the variables $U_{I,J+1/2,K\|L}$ and $U_{I,J+1/2,K\|R}$ at the cell interface $(I, J+1/2, L)$ as examples. Furthermore, to simplify the presentation, the subscript J will be omitted, and the variables will be referred to concisely as $U_{J+1/2\|L}$ and $U_{J+1/2\|R}$ respectively.

1. Zero-Order Reconstruction

The zeroth-order reconstruction has the advantage of being simple to construct and it imparts an upwind characteristic to the interface. However, it has the drawbacks of only having first-order accuracy and a weak capability for capturing shocks. The construction method is as follows:

$$U_{J+1/2\|L} = U_J, \quad U_{J+1/2\|R} = U_{J+1} \tag{2.42}$$

2. MUSCL Interpolation

MUSCL interpolation [10] can endow upwind schemes with monotonicity, and by adjusting parameters, it can enhance the accuracy of the upwind schemes. This interpolation can be expressed as:

$$\begin{aligned} U_{J+1/2\|L} &= U_J + \frac{1}{4}\left[(1-\kappa)\Delta_J^- + (1+\kappa)\Delta_J^+\right] \\ U_{J+1/2\|R} &= U_{J+1} - \frac{1}{4}\left[(1-\kappa)\Delta_{J+1}^+ + (1+\kappa)\Delta_{J+1}^-\right] \end{aligned} \tag{2.43}$$

where,

$$\Delta_J^+ = U_{J+1} - U_J \quad \Delta_J^- = U_J - U_{J-1} \tag{2.44}$$

The coefficient κ belongs to the interval $[-1, 1]$. When κ takes on different values, it constructs various finite difference schemes: when $\kappa = -1$ it results in a second-order upwind finite difference scheme; when $\kappa = 1/3$, it yields a third-order upwind finite difference scheme; and when $\kappa = 1$, it corresponds to a central finite difference scheme. In the central difference case, artificial viscosity must be introduced to prevent odd–even decoupling in smooth regions and to suppress numerical oscillations at discontinuities. However, it is important to note that for two-dimensional and three-dimensional flows, even when $\kappa = 1/3$, the scheme can only achieve second-order accuracy; third-order accuracy is attainable exclusively in one-dimensional flow scenarios.

3. MUSCL Interpolation Augmented with a Flux Limiter

In scenarios where the flow field exhibits discontinuities in motion, such as shock waves, the utilization of Eq. (2.43) may occasionally engender numerical oscillations. Consequently, it becomes imperative to employ MUSCL interpolation in tandem with a flux limiter. Flux limiters are capable of modulating higher-order terms in accordance with the spatial variations of the flow field parameters, thereby efficaciously averting the emergence of numerical oscillations throughout the computational process. Practical experience has shown that the use of flux limiters significantly affects stability and numerical accuracy. The judicious selection of a flux limiter, one that is congruent with the underlying discretization scheme, emerges as a critical determinant, exerting a pivotal influence on the stability, numerical precision, and convergence of the computational algorithm.

(1) MUSCL + The van Albada Limiter

The van Albada limiter [11] is one of the most commonly used flux limiters, which can effectively suppress numerical oscillations. The expression for the van Albada limiter in conjunction with MUSCL interpolation is as follows:

$$\begin{aligned} U_{J+1/2\|L} &= U_J + \frac{1}{4}\left[(1-\kappa s_J)\Delta_J^- + (1+\kappa s_J)\Delta_J^+\right] \\ U_{J+1/2\|R} &= U_{J+1} - \frac{1}{4}\left[(1-\kappa s_{J+1})\Delta_{J+1}^+ + (1+\kappa s_{J+1})\Delta_{J+1}^-\right] \end{aligned} \tag{2.45}$$

In which s represents the van Albada function, the expression is given by:

$$s_J = \frac{2\Delta_J^+\Delta_J + \varepsilon}{\left(\Delta_J^+\right)^2 + \left(\Delta_J^-\right)^2 + \varepsilon} \tag{2.46}$$

In the formula, ε is a small constant ($\varepsilon \approx 1 \times 10^{-6}$), which prevents s from being zero in smooth flow regions.

(2) MUSCL + The minmod Limiter

The minmod limiter [12] operates such that it takes the value of zero when the signs are different and the minimum absolute value when the signs are the same[5]:

$$\min\bmod(a,b) = \frac{\text{sign}(a) + \text{sign}(b)}{2}\min(|a|,|b|) \tag{2.47}$$

[5] The sign function is a mathematical function that returns the sign of a given number a, i.e., $\text{sign}(a) = \begin{cases} 1 & a > 0 \\ 0 & a = 0 \\ -1 & a < 0 \end{cases}$.

The expression for the minmod limiter combined with MUSCL interpolation is as follows:

$$\begin{aligned}
U_{J+1/2\|L} &= U_J + \frac{1}{4}\left[(1-\kappa)\min\bmod\left(\Delta_J^-, b\Delta_J^+\right) + (1+\kappa)\min\bmod\left(b\Delta_J^-, \Delta_J^+\right)\right] \\
U_{J+1/2\|R} &= U_J - \frac{1}{4}\left[(1-\kappa)\min\bmod\left(\Delta_{J+1}^-, b\Delta_{J+1}^+\right)\right. \\
&\quad \left. +(1+\kappa)\min\bmod\left(b\Delta_{J+1}^-, \Delta_{J+1}^+\right)\right]
\end{aligned} \tag{2.48}$$

where $1 \le b \le \frac{3-\kappa}{1-\kappa}$ with $b = 1$.

(3) The General Form of MUSCL Interpolation + Limiter

For the purpose of analytical comparison and to streamline the examination process, the various forms of flux limiters previously mentioned are consolidated into a singular, standardized formulation

$$\begin{aligned}
U_{J+1/2\|L} &= U_J + \frac{1}{4}\left[(1-\kappa)\varphi(r_L) + (1+\kappa)r_L\varphi\left(\frac{1}{r_L}\right)\right]_J \Delta_J^- \\
U_{J+1/2\|R} &= U_{J+1} + \frac{1}{4}\left[(1-\kappa)\varphi(r_R) + (1+\kappa)r_R\varphi\left(\frac{1}{r_R}\right)\right]_{J+1} \Delta_{J+1}^+
\end{aligned} \tag{2.49}$$

in which:

$$r_L = \frac{\Delta_J^+}{\Delta_J^-}, \quad r_R = \frac{\Delta_{J+1}^-}{\Delta_{J+1}^+}, \quad \text{it should be noted that } \Delta_J^+ = \Delta_{J+1}^-.$$

For different limiters, the value of $\varphi(r)$ varies:

$$\begin{aligned}
\text{van Albada limiter } \varphi(r) &= \frac{r+|r|}{1+r} \\
\text{minmod limiter } \varphi(r) &= \frac{r^2+r}{1+r^2}
\end{aligned} \tag{2.50}$$

Consequently, the functional paradigm of the flux limiter is thereby rendered consistent.

4. Utilizing Flux Limiters Solely

In the absence of MUSCL interpolation, employing flux limiters exclusively facilitates the attainment of first-order precision. The methodology for interpolation at the interface boundaries, under such circumstances, is delineated as follows:

(1) van Albada Limiter

The van Albada limiter formula is as follows:

$$U_{J+1/2\|L} = U_J - \frac{1}{2}S\left(\Delta_J^-, \Delta_J^+\right), \quad U_{J+1/2\|R} = U_{J+1} - \frac{1}{2}S\left(\Delta_J^-, \Delta_J^+\right) \tag{2.51}$$

In which $S(a, b) = \frac{(a^2+\varepsilon)b+(b+\varepsilon)a}{a^2+b^2+2\varepsilon}$.

(2) Minmod Limiter

The minmod limiter formula is as follows:

$$\begin{aligned} U_{J+1/2\|L} &= U_J - \frac{1}{2}\min\bmod\left(\Delta_J^-, \Delta_J^+\right), \\ U_{J+1/2\|R} &= U_{J+1} - \frac{1}{2}\min\bmod\left(\Delta_J^-, \Delta_J^+\right) \end{aligned} \tag{2.52}$$

5. Fifth-Order WENO Reconstruction

The Weighted Essentially Non-Oscillatory (WENO) scheme [13] is renowned for its stability, inherent non-oscillatory nature, and superior shock-capturing resolution, which have collectively contributed to its extensive consideration within the aerospace sector [14–16]. This methodology supports the construction of schemes ranging from third-order to seventh-order, and even higher, with the caveat that increased order of accuracy correlates with escalated computational demand. Striking a balance between computational intensity and precision, this text elects to elucidate the fifth-order WENO scheme as a moderate choice. The fifth-order WENO reconstruction entails leveraging the values from the five cell centers circumjacent to an interface—three upstream and two downstream. These points are then segregated into three stencils, each comprising three consecutive cell centers, as illustrated in Fig. 2.4. Upon each stencil, a third-order accurate interpolation is independently formulated. These interpolations are subsequently amalgamated through a weighting mechanism to attain a fifth-order accurate reconstruction.

When constructing the left-side variable $U_{J+1/2\|L}$, waves propagate from the left to the right, and the construction utilizes a left-biased set of five points, namely ($J-2, J-1, J, J+1, J+2$), as shown in Fig. 2.4a. The method for constructing the scheme is detailed in Eq. (2.53).

$$U_{J+1/2\|L} = \beta_1 q_{J+1/2}^{(1)} + \beta_2 q_{J+1/2}^{(2)} + \beta_3 q_{J+1/2}^{(3)} \tag{2.53}$$

in which:

$q_{J+1/2}^{(k)}$ the third-order interpolation of each stencil;

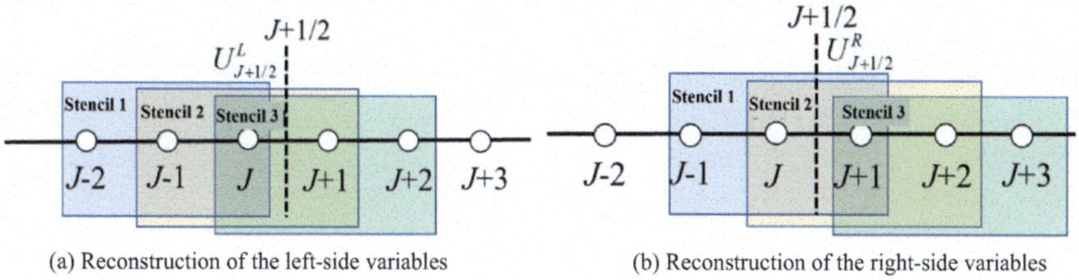

Fig. 2.4 Schematic diagram of the fifth-order WENO reconstruction for the variables on both sides of the interface

β_k the weight coefficients for each stencil.

The construction of the interpolation and weight coefficients for each stencil is detailed in Eq. (2.54).

$$
\begin{aligned}
& q^{(1)}_{J+1/2} = \frac{1}{3}U_{J-2} - \frac{7}{6}U_{J-1} + \frac{11}{6}U_J \\
& q^{(2)}_{J+1/2} = -\frac{1}{6}U_{J-1} + \frac{5}{6}U_J + \frac{1}{3}U_{J+1} \\
& q^{(3)}_{J+1/2} = \frac{1}{3}U_J + \frac{5}{6}U_{J+1} - \frac{1}{6}U_{J+2} \\
& \beta_k = \frac{\eta_k}{\eta_1 + \eta_2 + \eta_3}; \quad \eta_k = \frac{B_k}{(\varepsilon + IS_k)^e}; \quad k = 1, 2, 3; \quad e = 2; \quad \varepsilon = 10^{-6} \\
& B_1 = 1/10 \quad B_2 = 6/10 \quad B_3 = 3/10 \\
& IS_1 = \frac{1}{4}(U_{J-2} - 4U_{J-1} + 3U_J)^2 + \frac{13}{12}(U_{J-2} - 2U_{J-1} + U_J)^2 \\
& IS_2 = \frac{1}{4}(U_{J-1} - U_{J+1})^2 + \frac{13}{12}(U_{J-1} - 2U_J + U_{J+1})^2 \\
& IS_3 = \frac{1}{4}(3U_J - 4U_{J+1} + U_{J+2})^2 + \frac{13}{12}(U_J - 2U_{J+1} + U_{J+2})^2
\end{aligned}
\tag{2.54}
$$

The method for constructing the right-side variable $U_{J+1/2||R}$ is similar to the procedure described previously, with a key difference that pertains to the direction of wave propagation. In this instance, the waves are transmitted from the right to the left, which conceptually positions the right side as the upstream region. Consequently, the construction is executed using the points ($J - 1, J, J + 1, J + 2, J + 3$). This arrangement is reflectively symmetrical to the left-side construction format at the interface $J + 1/2$, as illustrated in Fig. 2.4b. The formulation for this construction is articulated in Eqs. (2.55) and (2.56).

$$
U_{J+1/2||R} = \beta_1 q^{(1)}_{J+1/2} + \beta_2 q^{(2)}_{J+1/2} + \beta_3 q^{(3)}_{J+1/2} \tag{2.55}
$$

$$
\begin{aligned}
&q_{J+1/2}^{(1)} = \frac{1}{3}U_{J+3} - \frac{7}{6}U_{J+2} + \frac{11}{6}U_{J+1} \\
&q_{J+1/2}^{(2)} = -\frac{1}{6}U_{J+2} + \frac{5}{6}U_{J+1} + \frac{1}{3}U_{J} \\
&q_{J+1/2}^{(3)} = \frac{1}{3}U_{J+1} + \frac{5}{6}U_{J} - \frac{1}{6}U_{J-1} \\
&\beta_k = \frac{\eta_k}{\eta_1 + \eta_2 + \eta_3}; \quad \eta_k = \frac{B_k}{(\varepsilon + IS_k)^{\xi}}; \quad k = 1, 2, 3; \quad e = 2; \quad \varepsilon = 10^{-6} \\
&B_1 = 1/10 \quad B_2 = 6/10 \quad B_3 = 3/10 \\
&IS_1 = \frac{1}{4}(U_{J+3} - 4U_{J+2} + 3U_{J+1})^2 + \frac{13}{12}(U_{J+3} - 2U_{J+2} + U_{J+1})^2 \\
&IS_2 = \frac{1}{4}(U_{J+2} - U_{J})^2 + \frac{13}{12}(U_{J+2} - 2U_{J+1} + U_{J})^2 \\
&IS_3 = \frac{1}{4}(3U_{J+1} - 4U_{J} + U_{J-1})^2 + \frac{13}{12}(U_{J+1} - 2U_{J} + U_{J-1})^2
\end{aligned}
\tag{2.56}
$$

The weight coefficients corresponding to each stencil are contingent upon the smoothness of the flow field. When addressing shock wave problems, the stability of the computation can be enhanced by adjusting the weight coefficients of the stencils in the vicinity of the shock. This adjustment effectively reduces the order of accuracy at the location of the shock, yielding a reduced third-order accuracy in the immediate vicinity of the shock waves.

2.3.3 Spatial Discretization Schemes on Non-uniform Grid

The WENO scheme is inherently designed for uniform grids, yet in engineering applications, non-uniform grids are often utilized for computations, which can diminish the accuracy of the fifth-order WENO scheme. This section investigates the impact of non-uniform grids on the precision of the scheme and discusses the necessity of preserving accuracy on non-uniform grids within the context of both the finite volume method and the finite difference method. The conclusions drawn may serve as a reference for researchers employing finite difference approaches.

The Impact of Non-uniform Grids on Accuracy

This section delves into the investigation of numerical scheme accuracy through the resolution of the initial value problem for a one-dimensional linear convection equation, as articulated in Eq. (2.57). The convection term is discretized utilizing a fifth-order WENO scheme, while the temporal component is discretized via a fourth-order R-K method, with a consistent time step applied across the whole filed. The grid system is tailored with unilateral refinement, as delineated in Fig. 2.5, to facilitate spatial discretization of the one-dimensional computational domain with Length 2π.

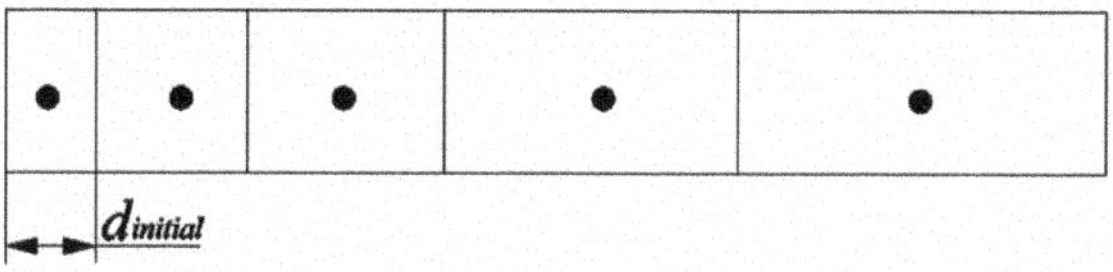

Fig. 2.5 Schematic diagram of a non-uniform mesh

The initial grid spacing of the most leftward cell is denoted as d_{initial}, with the total grid count specified as N_{cells}. When the computational domain is apportioned into a uniform grid, the resultant grid spacing is expressed as d_{average} ($d_{\text{average}} = 2\pi/N_{cells}$). The Grid Non-uniformity (GI) is quantified as $GI = d_{\text{average}} - d_{\text{initial}}/d_{\text{average}}$, with higher GI values indicative of increased grid non-uniformity. The study presents an in-depth analysis of the fifth-order WENO scheme's precision under conditions of grid non-uniformity, specifically examining scenarios with GI values of 20 and 80%. To ascertain that the temporal discretization does not compromise the spatial accuracy, the Courant–Friedrichs–Lewy (CFL) number is progressively diminished for each grid configuration until successive computational outcomes are found to be identical. The computational procedure is iterated over a duration equivalent to one unit of physical time.

$$\begin{cases} \frac{\partial g}{\partial t} + \frac{\partial g}{\partial x} = 0 & x \in [0, 2],\ \text{period},\ t > 0 \\ g(x, 0) = \sin(\pi x) & x \in [0, 2],\ \text{period} \end{cases} \tag{2.57}$$

The precision test results for non-uniform mesh formats when $GI = 20\%$ and $GI = 80\%$ are presented in Table 2.2. For comparison, the norm error (as Eq. (2.58)) and accuracy order of a uniform mesh are also provided. The results indicate that the uniform mesh ($GI = 0\%$) exhibits a 5th-order accuracy. However, when the mesh's GI is at 20%, the accuracy order is reduced to approximately 3rd-order, and when the GI increases to 80%, the accuracy is further reduced to around 2.5th-order. It is evident that as the degree of mesh non-uniformity increases, the accuracy order gradually decreases.

$$\begin{aligned} \|Error\|_1 &= \frac{1}{ncells} \sum_{i=1}^{ncells} |g_i - g(x_i)|; \\ order &= \ln\left(\|Error\|_1^m / \|Error\|_1^n\right) / \ln(ncells_n / ncells_m) \end{aligned} \tag{2.58}$$

where

g_i numerical computation results;
$ncells$ number of grids;
$g(x_i)$ analytical solution;
m, n the results from two sets of meshes; the results are typically selected from adjacent ones.

Table 2.2 The impact of mesh non-uniformity on accuracy

Uniform mesh ($GI = 0\%$)		$GI = 20\%$		$GI = 80\%$	
$\|\|Error\|\|_1$	Accuracy order	$\|\|Error\|\|_1$	Accuracy order	$\|\|Error\|\|_1$	Accuracy order
1.6462E−2	–	2.1929E−2	–	1.5444E−1	–
7.4923E−4	4.46	2.0082E−3	3.45	2.7223E−2	2.50
2.2870E−5	5.03	1.8259E−4	3.45	4.7377E−3	2.51
7.3131E−7	4.96	2.9600E−5	3.18	8.1919E−4	2.52
2.2638E−8	5.01	6.5823E−6	2.93	1.8537E−4	2.43

Analysis of the Necessity for Maintaining Accuracy

There are typically two methods for ensuring accuracy of WENO schemes on non-uniform meshes: ① Constructing high-precision WENO schemes on non-uniform meshes; ② Achieving precision preservation of the WENO scheme when used in a uniform computational coordinate system by transforming from a non-uniform physical coordinate system. There is an inherent contradiction between high precision and computational cost; whether to prioritize accuracy depends on the computational cost and the problem being studied, which will be discussed in detail below.

1. Finite Volume Method

In the Finite Volume Method, it is common to enhance accuracy by constructing discrete schemes on non-uniform meshes. For instance, Dutch researcher M. van Sint Annaland [17] and Canadian Raymond J. Spiteri [18] have both incorporated the mesh scale into the nonlinear weights of the WENO scheme to construct WENO formats on non-uniform meshes. Due to the complexity of construction, further details are not elaborated here and can be referred to in the literature [17, 18].

Since the construction of the WENO scheme on non-uniform meshes takes into account the local mesh scale, it undoubtedly increases the computational load. Taking the method from Ref. [17] as an example, each step of the WENO scheme adds approximately 280 arithmetic operations of addition/subtraction/multiplication/division. Tested on the Tianhe-2 platform, it was evaluated that for an area with 200,000 meshes, the CPU time per iteration is increased by about 1.43 s; for unsteady calculations, with 1000 physical time steps and 40 pseudo time steps per time step, the CPU time increases by 15.89 h. However, for high Reynolds number flow field problems such as compressors, the computational load is much greater than 200,000 meshes, usually at the level of millions. Therefore, using the Finite Volume Method to construct high-precision WENO schemes on non-uniform meshes is computationally prohibitive. On the other hand, when using DES class methods for computation, refining the mesh in key areas with complex vortex structures helps to reduce numerical errors.

2. Finite Difference Method

In the Finite Difference Method, the common practice is to employ a coordinate transformation technique to convert the derivatives of convective terms from a non-uniform physical coordinate system into derivatives within a uniform computational coordinate system. Specifically, when considering a one-dimensional scalar conservation Eq. (2.57), the chain rule is applied within a non-uniform mesh to construct a finite difference scheme at grid node i: $\mathrm{d}g_i/\mathrm{d}t + \left(g_{i+1/2} - {}_{i-1/2}\right)/\left(x_\xi\right)_i = 0$. Here, ξ represents the computational coordinate, and regardless of whether the mesh is uniform in the physical coordinate system, the mesh within the computational coordinate system is a uniform mesh. x_ξ is the derivative of the physical coordinate with respect to the computational coordinate, and three methods are adopted to handle this derivative:

(1) Centered Difference (abbreviated as Center):

$$x_{j+1/2} = \frac{1}{2}\left(x_{j+1} + x_j\right) \tag{2.59}$$

The centered difference scheme theoretically offers second-order accuracy. The accuracy of the coordinate derivative is lower than the discretization accuracy of the numerical flux, which is referred to as a lower-order treatment method.

(2) Fifth-Order Weighted Essentially Non-Oscillatory (abbreviated as WENO5):

$$x_{j+1/2} = \omega_1 x_{j+1/2}^{(1)} + \omega_2 x_{j+1/2}^{(2)} + \omega_3 x_{j+1/2}^{(3)} \tag{2.60}$$

For a detailed introduction, see Sect. 2.3.2. At this point, the accuracy of the coordinate derivative matches the accuracy of the numerical flux, which is referred to as a consistent order treatment method.

(3) Seventh-Order Weighted Essentially Non-Oscillatory (abbreviated as WENO7):

$$x_{j+1/2} = \omega_1 x_{j+1/2}^{(1)} + \omega_2 x_{j+1/2}^{(2)} + \omega_3 x_{j+1/2}^{(3)} + \omega_4 x_{j+1/2}^{(4)} \tag{2.61}$$

The weight coefficients ω_k and the computation of numerical fluxes for different stencils $x_{j+1/2}^k$ can be found in Ref. [19]. At this point, the precision of the coordinate derivative exceeds the precision of the numerical flux, which is referred as a high-order treatment method.

To evaluate the numerical accuracy of different coordinate transformation treatments in a non-uniform coordinate system, the results are depicted in Fig. 2.6, showing the outcome after the mesh number has increased to the point where accuracy no longer changes. It can be observed that employing a lower-order treatment method (Center) leads to a reduction in accuracy to approximately second-order. A consistent order treatment method (WENO5) brings the WENO scheme's accuracy close to fourth-order, while utilizing a higher-order treatment method (WENO7) can achieve accuracy above fourth-order.

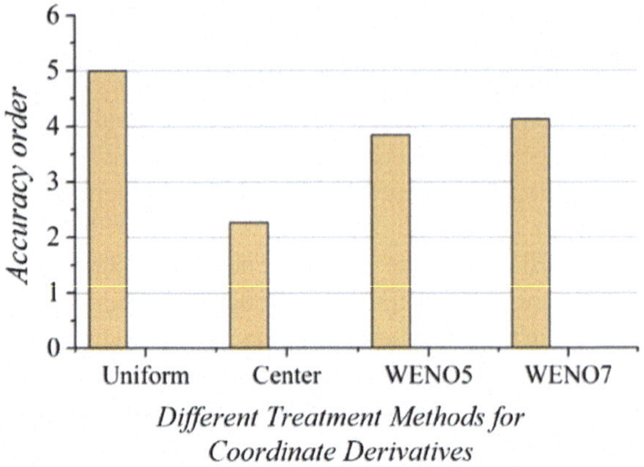

Fig. 2.6 Accuracy of the fifth-order WENO scheme with different treatments of coordinate derivatives

In summary, for the Finite Difference Method, adopting consistent or higher-order treatment methods for the coordinate derivatives on non-uniform meshes can effectively ensure the accuracy of the schemes on such meshes. Moreover, for problems where the mesh does not change over time, this coordinate transformation derivatives treatment is calculated only once throughout the entire solution period, accounting for a very small proportion. Therefore, for the Finite Difference Method, it is necessary to undertake accuracy preservation treatment on non-uniform meshes.

2.4 Convergence Criterion

1. Steady Calculation Convergence Criterion

In steady calculations, the following three criteria are mainly used to determine whether the calculation converges:

(1) The total residual root mean square (RMS) is defined as the mean square value of the change of flow parameters in the entire field. When the total residual in the entire field drops to more than four orders of magnitude, iteration can be considered to have converged;
(2) The error between inlet and outlet flow is controlled within 1.0%;
(3) As the number of iterations increases, the total performance parameters such as flow rate, pressure ratio, efficiency, etc. do not change with the increase in the number of iteration steps; in addition, parameters such as pressure or velocity at a certain monitoring point in the flow field remain basically unchanged.

2. Unsteady Calculation Convergence Criterion

For unsteady calculations, since the flow field changes with time, the above-mentioned convergence criteria used in steady calculations are no longer applicable. Due to the unsteady interference characteristics between relatively moving parts in the same flow field, when performing unsteady calculations, the inlet and outlet flow rates fluctuate periodically, and the changes between the two are inconsistent. Therefore, unsteady calculations are performed by monitoring the total performance parameters and the flow rate. The periodicity of pressure changes at a given point in the field. When the total performance parameters and the pressure signal of the monitoring point change periodically with time, the unsteady calculation is considered to have converged. After the calculation converges, all flow field parameters change periodically with time, and the period is the time for adjacent blade rows to rotate through one grid pitch.

2.5 DES Series Methods

The DES series methods represent an amalgamative strategy that integrates the computational frameworks of RANS (Reynolds-Averaged Navier–Stokes) and LES (Large Eddy Simulation). Within the periphery of the boundary layer, the RANS approach is leveraged for its computational efficiency. Extrapolating beyond the boundary layer, the LES technique is deployed to explicitly simulate the large-scale turbulent eddies. The quintessence of the DES methodology entails an adjustment to the turbulent hybrid length within the RANS model's turbulence framework, complemented by the integration of a model switch mechanism. This mechanism is instrumental in delineating the operational zones for both RANS and LES approaches, thereby facilitating a dynamic and adaptive computational strategy.

The purpose of a turbulence model is to determine the turbulent viscosity coefficient that appears in the Reynolds stress model, thereby achieving closure of the governing equations. Turbulence models within the framework of the Reynolds-Averaged Navier–Stokes (RANS) approach are generally categorized into eddy viscosity models based on the Boussinesq hypothesis[6] and Reynolds stress-equation models (RSM) constructed through the Reynolds stress transport equation. Due to considerations of computational cost and convergence, eddy viscosity models are more commonly used. These eddy viscosity-based turbulence models can be further divided into zero-equation models (e.g., Baldwin–Lomax, B-L), one-equation models (e.g., Spalart–Allmaras, S-A), and two-equation models (e.g., SST, standard $k-\varepsilon$, standard $k-\omega$).

Among these various turbulence models, the S-A model is particularly favored for its low computational demand, good performance in flows with adverse pressure gradients, and suitability for high-speed flows. It is primarily applied in the numerical simulation of the flow fields of supersonic and transonic aeroengines. Based on the research group's extensive experience in predicting compressor flow fields [20], the S-A model has demonstrated excellent predictive capabilities for time-averaged

[6] Boussinesq Hypothesis: The Reynolds stress is proportional to the mean velocity gradient, with the proportionality coefficient being the turbulent viscosity. This assumption is reasonable for simple shear flows, such as boundary layers, pipe flows, and round jets.

parameters under stable operating conditions. Therefore, this book introduces the development of DES-type methods based on the S-A model.

2.5.1 Spalart–Allmaras Turbulence Model

Introduced in 1992 by Spalart and Allmaras [21], the S-A model posits that the turbulent viscosity coefficient adheres to a scalar transport equation that is applicable across various flow fields [22]. The model delineates the mathematical formulations for its constituent terms through empirical postulates and dimensional analysis. Extensive aerodynamic experiments on airfoils have been utilized to refine the empirical parameters, thereby establishing the S-A model as a prevalent turbulence model within the aerospace sector.

The universal scalar transport equation encapsulates the intricate interplay between the principal mechanisms of advection, diffusion, production, and dissipation within the flow field. The full turbulence S-A model for the kinematic viscosity coefficient is articulated in the conservation form of Eq. (2.62), wherein $\hat{\upsilon}_T$ denotes the pseudo-viscosity coefficient and $\rho\hat{\upsilon}_T$ signifies the conserved variable derived from the turbulence model equation. By conceptualizing the components on the right-hand side of the equation as source terms, Eq. (2.62) mirrors the structural form of the governing Eq. (2.1). Thus, it is feasible to integrate the full turbulence S-A model, as delineated by Eq. (2.62), with the N-S equations, as represented by Eq. (2.1), for a coupled solution.

$$\frac{\partial}{\partial t}(\hat{\upsilon}_T) + \overbrace{\frac{\partial}{\partial x_i}(\hat{\upsilon}_T w_i)}^{\text{advection}} = \underbrace{\frac{1}{\sigma}\left\{\frac{\partial}{\partial x_i}\left[\left(\frac{\mu_l}{\rho} + \hat{\upsilon}_T\right)\frac{\partial \hat{\upsilon}_T}{\partial x_i}\right] + c_{b2}\frac{\partial \hat{\upsilon}_T}{\partial x_i}\frac{\partial \hat{\upsilon}_T}{\partial x_i}\right\}}_{\text{diffusion}}$$
$$+ \underbrace{c_{b1}(1 - f_{t2})\hat{S}\hat{\upsilon}_T}_{\text{production}} - \underbrace{\left(c_{w1}f_w - \frac{c_{b1}}{\kappa^2}f_{t2}\right)\left(\frac{\hat{\upsilon}_T}{d}\right)^2}_{\text{dissipation}} \tag{2.62}$$

The relationship between the turbulent viscosity coefficient and the pseudo-viscosity coefficient is given by:

$$\begin{aligned} \mu_t &= \rho\hat{\upsilon}_T f_{v1} \\ f_{v1} &= \frac{\chi^3}{\chi^3 + c_{v1}^3} \quad \chi = \frac{\hat{\upsilon}_T}{\mu_l/\rho} \end{aligned} \tag{2.63}$$

Its dimensionless form is given by:

$$\frac{\partial}{\partial t}(\hat{\upsilon}_T) + \frac{\partial}{\partial x_i}(\hat{\upsilon}_T w_i) = \frac{1}{Re_{ref}}\frac{1}{\sigma}\left\{\frac{\partial}{\partial x_i}\left[\left(\frac{\mu_l}{\rho} + \hat{\upsilon}_T\right)\frac{\partial \hat{\upsilon}_T}{\partial x_i}\right] + c_{b2}\frac{\partial \hat{\upsilon}_T}{\partial x_i}\frac{\partial \hat{\upsilon}_T}{\partial x_i}\right\}$$
$$+ c_{b1}(1 - f_{t2})\hat{S}\hat{\upsilon}_T - \frac{1}{Re_{ref}}(c_{w1}f_w - \frac{c_{b1}}{\kappa^2}f_{t2})\left(\frac{\hat{\upsilon}_T}{d}\right)^2 \quad (2.64)$$

In Eq. (2.64), the term "*d*" within the dissipation item represents the distance from the computational point to the nearest wall surface, also known as the "hybrid length." The expressions for the production term and the remaining variables in the dissipation term are as follows [23]:

$$\text{production:} \quad \hat{S} = S + \frac{1}{Re_{ref}}\frac{\hat{\upsilon}_T}{\kappa^2 d^2}f_{v2}, f_{t2} = c_{t3}e^{(-c_{t4}\chi^2)} \quad (2.65)$$

In which: $S = \sqrt{2\psi_{ij}\psi_{ij}}$, $\psi_{ij} = \frac{1}{2}\left(\frac{\partial w_i}{\partial x_j} - \frac{\partial w_j}{\partial x_i}\right)$, $f_{v2} = 1 - \frac{\chi}{1+\chi f_{v1}}$

$$\text{dissipation:} \quad f_w = g\left[\frac{1 + c_{w3}^6}{g^6 + c_{w3}^6}\right]^{1/6} \quad \text{where} \quad g = r + c_{w2}(r^6 - r);$$
$$r = \frac{1}{Re_{ref}}\frac{\hat{\upsilon}_T}{\hat{S}\kappa^2 d^2} \quad (2.66)$$

the constants involved in Eq. (2.64) are as follows:

$$\sigma = 2/3, \quad c_{b2} = 0.622, \quad c_{b1} = 0.1355, \quad c_{w1} = \frac{c_{b1}}{\kappa^2} + \frac{1 + c_{b2}}{\sigma},$$
$$c_{w2} = 0.3, \quad \kappa = 0.41, \quad c_{w3} = 2, \quad c_{t3} = 1.2, \quad c_{t4} = 0.5 \quad (2.67)$$

Ensuring computational precision necessitates the optimization of the grid topology in the vicinity of the wall surface. Within the S-A model, the variable "*d*" signifies the spatial extent from the grid center to the nearest wall surface, as illustrated by the blue dashed line in Fig. 2.7. However, within the discrete spatial domain, precisely ascertaining the perpendicular distance to the nearest wall is non-trivial. Therefore, the linear distance to the closest discrete center point adjacent to the wall is typically utilized, as indicated by the red dashed line in Fig. 2.7. This approximation can engender discrepancies in the computational outcomes of the turbulence model. The turbulent viscosity coefficient typically plays a dominant role in regions close to the wall, hence to ensure the accuracy of the model's application, it is essential to maintain the orthogonality of the near-wall grid with respect to the wall surface.

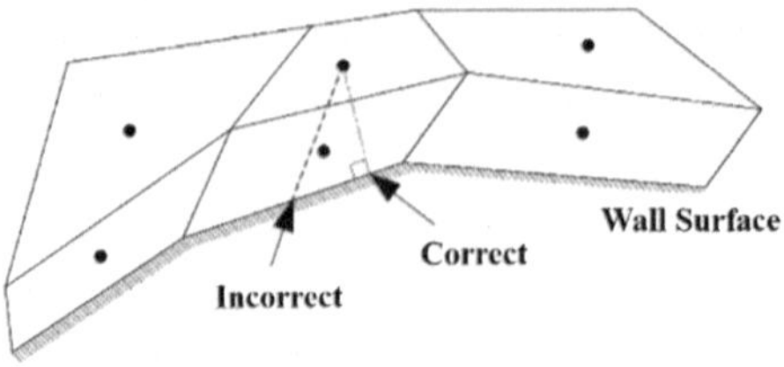

Fig. 2.7 The distance to the nearest wall surface in the S-A model

2.5.2 DES Method and Gray Area Issues

The DES method [22] modifies only the dissipation term of the S-A model, specifically by transforming the hybrid length d in the transport equation's dissipation term to $\tilde{d}_{\text{DES}}$, where $\tilde{d}_{\text{DES}}$ serves as the model switch function. This modification correlates the computation of the flow field with the grid scale. Taking the grid point at location k as an example:

$$\left(\tilde{d}_{\text{DES}}\right)_k = \min(d_k, C_{\text{DES}}\Delta_k) \tag{2.68}$$

In which:

C_{DES} empirical coefficient, commonly taken as 0.65;
Δ_k the filtering scale of the LES subgrid-scale model is chosen to be the maximum scale of the cell, i.e., $\Delta_k = \max(\Delta x_k, \Delta y_k, \Delta z_k)$.

In the schematic diagram of the DES method, Fig. 2.8, the four grids numbered $k = 1, 2, 3, 4$ have the closest distances to the surface denoted as d_k. For the computational points near the surface, $k = 1, 2$, it holds that $d_k < C_{\text{DES}}\Delta_k$. In accordance with Eq. (2.68), this yields $\left(\tilde{d}_{\text{DES}}\right)_k = d_k$ for the indices $k = 1, 2$. $\left(\tilde{d}_{\text{DES}}\right)_k = d_k$. At this point, the DES method's dissipation term is consistent with the original turbulence model (2.62), effectively behaving as a RANS method. From a physical perspective, within the boundary layer close to the wall, the small-scale vortex structures still satisfy the isotropy hypothesis, thereby validating the efficacy of the S-A turbulence model. Conversely, for computational nodes situated at a greater remove from the surface, for $k = 3, 4$, the condition $d_k > C_{DES}\Delta_k$ consequently leading to $\left(\tilde{d}_{\text{DES}}\right)_k = C_{\text{DES}}\Delta_k$. When the production term and the dissipation term in the transport equation (Eq. 2.62) reach equilibrium, we have:

$$c_{b1}\hat{S}\hat{\upsilon}_T = c_{w1}f_w\left(\frac{\hat{\upsilon}_T}{C_{DES}\Delta_k}\right)^2 \tag{2.69}$$

it follows that $\hat{\upsilon}_T \propto \hat{S}(C_{\text{DES}}\Delta_k)^2$, and upon the attainment of a state of local equilibrium, the coefficient of eddy viscosity is found to be directly proportional to the square of the vorticity magnitude and the subgrid scale, a criterion that is congruent with the requirement of the Smagorinsky eddy viscosity model of LES. Consequently, the DES approach effectively implements the implicit filtering characteristic of LES.

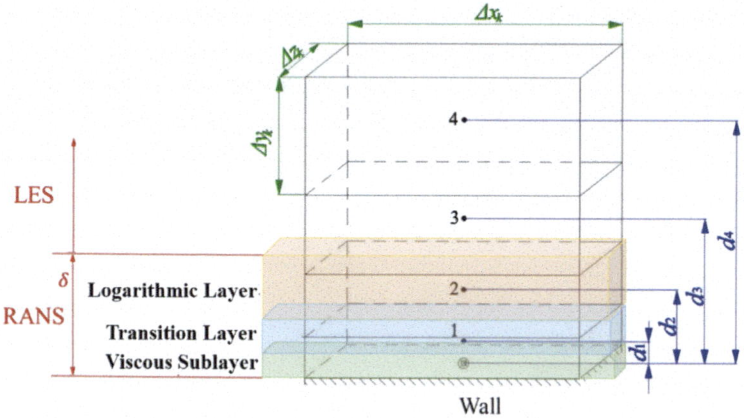

Fig. 2.8 Schematic diagram of DES regions (δ boundary layer thickness; d_k the distance from the grid center point to the nearest wall; Δx_k, Δy_k, Δz_k grid cell size)

The original intention and expected pattern of DES97 [22] are shown in Fig. 2.9a. Users are expected to partition the grid based on the characteristics of the steady-state flow field. The grid size will determine the regions of different turbulence methods. A reasonable RANS/LES region can correspond to the flow field characteristics to achieve the effect of using RANS near the wall and LES away from the wall. However, in practice, the selection of the characteristic length $\tilde{d}_{DES}$ in the DES97 method is solely contingent upon the local grid size. This dependency can engender substantial variations in the delineation of RANS and LES regions within an identical flow field when subjected to different grid partitioning strategies. For any set of grids, the position close to the wall is definitely "modeled" by RANS, and the position far from the wall is definitely "resolved" by LES. However, in the transitional zone proximate to the interface of these two domains, alterations in the grid scale—attributed to the method of partitioning—can exert a pronounced influence on the selection of the pertinent turbulence modeling approach. This transitional zone is referred to as the "gray area." According to the foundational principles upon which the DES method is constructed, the existence of a "gray area" is an unavoidable consequence.

The ideal existence of the gray area, as depicted in Fig. 2.9b, is positioned exterior to the boundary layer and is comparatively slender along the normal direction to the wall, ensuring a swift transition from the RANS region to the LES region. However, when the geometry of the object under investigation is complex, the configuration of the wall-adjacent boundary layer and the refinement of the grid near the wall also become correspondingly intricate, leading to alterations in the manifestation of the gray area. This situation gives rise to two types of "gray area issues."

The first category of gray area typically occurs with the refinement of near-wall grids, particularly when the Δx or Δz are extremely refined as shown in Fig. 2.9c. As a result, the transition switch from RANS to LES will be activated prematurely at the grid located in the outer layer of the boundary layer where $C_{DES}\Delta < d$. The

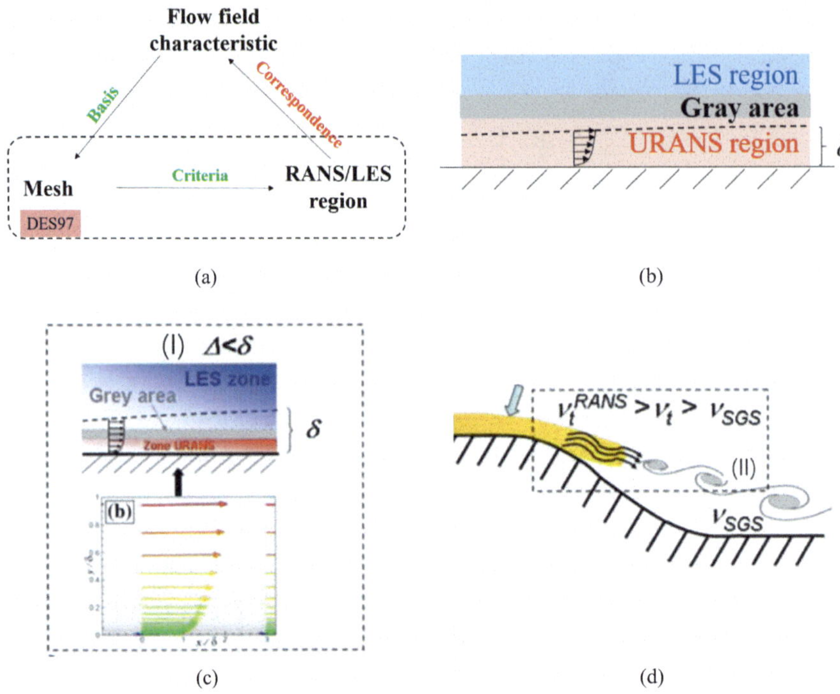

Fig. 2.9 Diagram of the gray area issue of the DES method. **a** The original intention of the DES method. **b** Ideal existence of the gray area. **c** First type "gray area issue" [23]. **d** Second type "gray area issue" [24]

grid size in this region is far from meeting the LES criteria for resolving the velocity fluctuations within the boundary layer, leading to modeled stress depletion (MSD) and grid induced separation (GIS) [23]. This arises precisely because the DES97 method's judgment between RANS and LES heavily relies on the local grid size, neglecting the local physical flow characteristics.

The second category of gray areas issues typically arises in geometrically induced shear flows. Figure 2.9d illustrates an example of a shear flow over a backward-facing step. In this model, as the airflow exits the region I adjacent to the wall, the velocity of the airflow on the lower side decreases due to an expansion in the cross-sectional area. There is a disparity in the airflow along the x-direction on either side of the interface, which results in shear action and the formation of an initial shear layer.

From the perspective of flow characteristics, minor disturbances near the initial shear layer can lead to the Kelvin–Helmholtz (K-H) instability. The development of K-H instability causes an intensification of the deformation within the shear layer, exhibiting a positive feedback mechanism. Consequently, this phenomenon can engender significant perturbations within the flow field, which in turn imparts a substantive impact on the evolution of the downstream flow dynamics.

From the standpoint of numerical simulation, an accurate resolution of the K-H instability at the initial shear location is a prerequisite for correctly predicting the flow field, which under ideal conditions should swiftly transition to LES [25]. However, in practice, for the wall region II, the development of the boundary layer is aligned with the direction of jet evolution and the interaction with the shear layer, that is, the x-direction. Therefore, a gray area exists in this direction. Regarding the upstream near-wall region, which is within the RANS modeling domain, it influences the viscosity of the downstream through convective effects. Under the combined interference of both, cannot transition rapidly from the RANS domain to the LES domain. This results in the Slow LES Development (SLD) issue, the second category of gray area problems.

Consequently, the advancement of DES series methods is not accomplished in a single iteration but is instead a progressive refinement process, addressing the issues that emerge at various stages. For the first type of gray zone issue, the core idea is to ensure the use of RANS within the boundary layer and LES outside of it. For the second type of gray area issue, the core idea is to accelerate the transition from RANS to LES in the initial region of shear layer development. Subsequently, we will detail several primary strategies for addressing the gray area issues.

2.5.3 Improved DES Methods for Gray Area Issues

The fundamental cause of the two types of gray area issues lies in the DES97 method's excessive reliance on grid size for the partitioning between RANS and LES, neglecting local flow characteristics. Spalart and Shur proposed the DDES [26] (Delayed Detached Eddy Simulation) and EDDES (Enhanced Delayed Detached Eddy Simulation) [27] methods, respectively. Both methods are global DES improved Method, corresponding to the first and second types of gray area issues. The improvement idea is to introduce a correction function throughout the entire flow field, adaptively modifying the characteristic length $\tilde{d}_{DES}$ for each grid based on local flow field features, making its eddy viscosity coefficient suitable for a reasonable turbulence method at this location. The subsequent exposition delves into the methodologies of DDES and EDDES, both of which are formulated within the paradigm of the S-A turbulence model.

DDES Method

Spalart proposed the DDES method in 2006 [26], where the modified hybrid length is denoted by $\tilde{d}_{DDES}$. Similarly, taking the hybrid length at grid cell k as an example, to simplify the expression, the subscript k is omitted without causing ambiguity.

$$\tilde{d}_{\mathrm{DDES}} = d - f_d \max(0, d - C_{\mathrm{DES}}\Delta) \tag{2.70}$$

$$f_d = 1 - \tanh\left([8r_d]^3\right) \tag{2.71}$$

Replacing r in Eq. (2.66) with r_d, which is defined similarly to r:

$$r_d = \frac{\hat{\upsilon}_T + \mu_l/\rho}{\sqrt{L_{i,j}L_{i,j}\kappa^2 d^2}} \tag{2.72}$$

in which:

$L_{i,j}$ velocity gradient.

In the expression for r_d, the inclusion of the dynamic viscosity coefficient μ_l in the numerator ensures a non-vanishing value of r_d in proximity to the wall. As delineated in Eqs. (2.27) through (2.45), the transition from the RANS to the LES domain is not discrete but rather a continuous spectrum, modulated by the coefficient f_d, thereby enabling a smoother interplay between the two solution regimes.

When $d > C_{\text{DES}}\Delta$ and $0 < f_d < 1$, at this point $C_{\text{DES}}\Delta < \tilde{d}_{\text{DDES},k} < d$ and the hybrid length scale $\tilde{d}_{\text{DDES}}$ is no longer entirely determined by the grid size. Instead, it is jointly determined by the grid size Δ, the distance to the wall d, and the velocity gradient of the flow field $L_{i,j}$.

In this manner, the interface between RANS and LES within the DES method is not static but varies with the flow field as the iterative calculations progress, effectively reducing the method's dependency on the grid. Owing to the superior performance of the DDES approach, in 2009, Spalart proposed that DDES be adopted as the new standard DES method [28].

EDDES Method

In response to this issue, Shur [27] proposed an Enhanced Delayed Detached Eddy Simulation (EDDES) method based on the Vortex Tilting Measure (VTM). This approach reduces eddy viscosity at the location where shear layer separation begins, thereby promoting the instability and development of the shear layer.

Initially, it modifies the sub-grid scale in the original DDES method by [29]:

$$\tilde{\Delta}_\omega = \frac{1}{\sqrt{3}} \max_{n,m=1,8} |\vec{n}_\omega \times (\vec{r}_n - \vec{r}_m)|$$

where $\vec{n}_\omega$ represents the unit vector along the local vorticity axis, and $\vec{r}_n$ represents the displacement vector between any two grid vertices. This can reduce the characteristic scale of the corresponding grid at the quasi-two-dimensional shear location during the initial separation. As a result, it reduces eddy viscosity, promoting the instability of the shear layer. Furthermore, Shur multiply the new sub-grid scale by a flow-field detection function capable of distinguishing between quasi-two-dimensional shear and fully developed three-dimensional turbulence. The detection function is dependent on the VTM sensor. Its expression is as follows:

$$F_{KH}(VTM) = \max\left\{F_{KH}^{\min}, \min\left[F_{KH}^{\max}, F_{KH}^{\min} + \frac{F_{KH}^{\max} - F_{KH}^{\min}}{a_2 - a_1}(VTM - a_1)\right]\right\}$$

$$VTM = \frac{\sqrt{6}\left|(\hat{S}\cdot\vec{\omega})\times\vec{\omega}\right|}{\omega^2\sqrt{3\mathrm{tr}(\hat{S}^2) - \left[\mathrm{tr}(\hat{S})\right]^2}}$$

where the coefficients $F_{KH}^{\max}, F_{KH}^{\min}, a_2, a_1$ are equal to $1, 0.1, 0.3$, and 0.1, respectively. When the eigenvalue tensor of the shear strain rate tensor $\hat{S}$ is correlated with the local vorticity vector, i.e. $\vec{\omega}$, fluid deformation occurs in a plane perpendicular to the local vorticity vector, the $\left|(\hat{S}\cdot\vec{\omega})\times\vec{\omega}\right|$ approaches 0 while the VTM value approaches 0. This indicates that the fluid deformation at this position is quasi-two-dimensional, implying it is at the initial shear location. On the contrary, when the fluid is in a fully developed three-dimensional turbulence, $\hat{S}$ is uncorrelated with the local vorticity vector $\vec{\omega}$, and the value of the VTM approaches 1.

Finally, by using a piece-wise function, $F_{KH}(VTM)$, it achieves a sub-grid scale close to $0.1\tilde{\Delta}_{\omega}$ in the initial shear region and the original sub-grid scale in the region of fully developed turbulence. This achieves the purpose of reducing eddy viscosity in the initial shear region while ensuring that in the region of fully developed turbulence remains at a normal level.

ZDES Method

Addressing the gray area issues, Deck diverged from the conventional global DES approaches by introducing a novel conceptual framework in 2005 [30], known as the Zonal Detached Eddy Simulation (ZDES) method. This methodology delineates distinct computational domains for RANS and DES, thereby ensuring that the boundary layer is exhaustively resolved through RANS modeling. This strategy effectively mitigates the first type gray area issue. Nonetheless, the inaugural rendition of ZDES necessitated the manual demarcation of the RANS and DES domains, which was contingent upon prior expertise and pre-existing flow field insights.

In 2012, drawing inspiration from the DDES methodology, Deck enhanced the ZDES approach by incorporating a delay function [31]. The revised ZDES categorizes flow separation into three primary types, as depicted in Fig. 2.10: (a) separation induced by geometric discontinuities; (b) separation due to variations in wall curvature; and (c) separation triggered by the dynamics of incoming boundary layer. This iteration of the ZDES method permits users to employ RANS or various DES modes within distinct flow characteristic regions of the same flow field, thereby affording greater flexibility for computational simulations.

The ZDES method is implemented on the basis of multiple grid blocks, achieving the objective of employing distinct turbulence methods in various regions of the same flow field by defining different grid blocks. Within a flow field composed of N grid

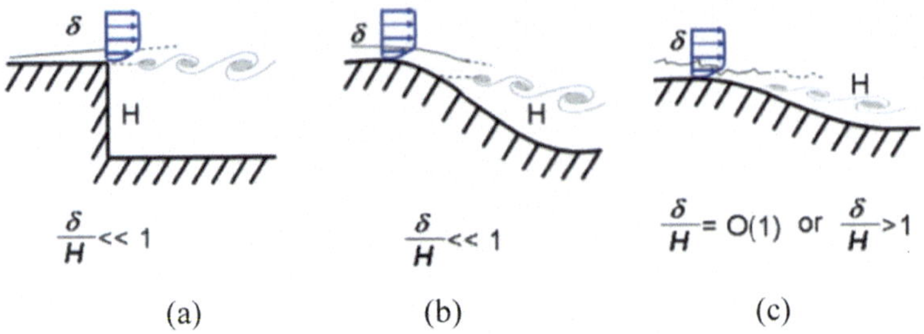

Fig. 2.10 Classification of typical flow problems [31]: **a** separation fixed by the geometry, **b** separation induced by a pressure gradient on a curved surface, **c** separation strongly influenced by the dynamics of the incoming boundary layer

blocks, for the nth grid block $(0 < n < N)$, when $ides_n = 0$, the RANS method is utilized in that region; when $ides_n = 1$, the DES method is applied. Addressing the three distinct separation problems depicted in Fig. 2.10, different hybrid lengths are employed, as illustrated by Eq. (2.73).

$$\tilde{d}_{\mathrm{ZDES}} = \begin{cases} d_{\mathrm{RANS}} & \text{if ides} = 0 \\ \tilde{d}_{\mathrm{DES}}^{\mathrm{I}} & \text{if ides} = 1 \text{ and imode} = 1 \\ \tilde{d}_{\mathrm{DES}}^{\mathrm{II}} & \text{if ides} = 1 \text{ and imode} = 2 \\ \tilde{d}_{\mathrm{DES}}^{\mathrm{III}} & \text{if ides} = 1 \text{ and imode} = 3 \end{cases} \tag{2.73}$$

When $ides_n = 1$ and imode $= 1$, the region corresponds to the Type I flow category, which is characterized by separation induced by geometric configuration discontinuities:

$$\tilde{d}_{\mathrm{DES}}^{\mathrm{I}} = \min\left(d, C_{\mathrm{DES}}\tilde{\Delta}_{\mathrm{DES}}^{\mathrm{I}}\right) \tag{2.74}$$

When $ides_n = 1$ and imode $= 2$, the region corresponds to the Type II flow category, which involves separation caused by adverse pressure gradients due to wall curvature, with the location of separation being a priori unknown.

$$\tilde{d}_{\mathrm{DES}}^{\mathrm{II}} = d - f_d \max\left(0, d - C_{\mathrm{DES}}\tilde{\Delta}_{\mathrm{DES}}^{\mathrm{II}}\right) \tag{2.75}$$

When $ides_n = 1$ and imode $= 3$, this region corresponds to the Type III flow category, which involves separation caused by fluctuations in the incoming boundary layer.

$$\tilde{d}_{\mathrm{DES}}^{\mathrm{III}} = \begin{cases} d_{\mathrm{w}} & \text{if } d_{\mathrm{w}} < d_{\mathrm{w}}^{\mathrm{interface}} \\ \tilde{d}_{\mathrm{DES}}^{\mathrm{I}} & \text{otherwise} \end{cases} \tag{2.76}$$

In which, $C_{\mathrm{DES}} = 0.65$, and f_d is the delay function, for details see Sect. 2.5.3.1. From Eqs. (2.73) to (2.76), it can be observed that the ZDES method not only offers zonal flexibility in different locations but also redefines the DES characteristic length for the three types of separation issues to address gray area problems. This paper primarily focuses on separations originating from the first two categories, hence an in-depth introduction to the third type of separation is not provided.

$$\Delta_{\mathrm{DES}}^{\mathrm{I}} = \Delta_{\omega}$$
$$\Delta_{\mathrm{DES}}^{\mathrm{II}} = \begin{cases} \Delta_{\max}, & \text{if } f_d < f_{d0} \\ \Delta_{\omega}, & \text{if } f_d > f_{d0} \end{cases} \tag{2.77}$$

In which, $\Delta_{\max}$ represents the original grid scale definition in DES97, and Δ_{ω} is a new subgrid scale introduced by Chauvent, specifically addressing the second type of gray area issue, namely the SLD problem. Its definition is presented in Eq. (2.78).

$$\Delta = \Delta_{\omega} = \sqrt{N_x^2 \Delta y \Delta z + N_y^2 \Delta z \Delta x + N_z^2 \Delta x \Delta y} \tag{2.78}$$

Δ_{ω} characterizes a subgrid scale based on local vorticity weighting, where N_x, N_y, N_z are the three components of the vorticity unit vector $\boldsymbol{N} = \boldsymbol{\omega}/\|\boldsymbol{\omega}\|$ at the grid center node. In the initial shear region, $N_z >> N_x \ or \ N_y$, resulting $\Delta_{\omega} = \sqrt{\Delta x \Delta y}$. On the contrary, in the fully developed three-dimensional isotropic turbulence region, $N_z \approx N_x \approx N_y$, and at this point, Δ_{ω} is close to the original DES97 definition. It has validated that the accuracy of Δ_{ω} when applied to the SLD problem [29], thus it can be directly substituted for the original DES characteristic length in the first category of separations caused by geometry.

In the case of the second type of separation, employing Δ_{ω} directly within the boundary layer would reduce the eddy viscosity, potentially exacerbating the first type of gray area issue. Consequently, within the boundary layer, $\Delta_{\max}$ is utilized, and at the outer layer of the boundary where a rapid transition is required, the delay function f_d is employed for the rapid shift from $\Delta_{\max}$ to Δ_{ω}. Additionally, f_{d0} is introduced to denote the outer boundary of the boundary layer, with a value range of (0.75, 0.99), ensuring that the original DDES behavior is maintained within the boundary layer.

2.5.4 *IDDES Method*

In 2000, Nikitin introduced the DES as a Wall-Modeled Large Eddy Simulation (WMLES) approach [32], employing RANS within the inner boundary layer and LES in the outer boundary layer. However, this method was associated with a deviation in the velocity profile known as "Log-Layer Mismatch" (LLM), which led to an underprediction of skin friction by approximately 15% [32], rendering it inaccurate. To address this issue, Shur proposed the Improved Delayed Detached Eddy Simulation (IDDES) method [33], which redefines the hybrid length scale $\tilde{d}_{\mathrm{IDDES}}$, as shown in Eq. (2.79). This redefinition effectively blends the filtering scale d used in the RANS method with the filtering scale Δ used in the LES method.

$$\tilde{d}_{\mathrm{IDDES}} = f_{hyb}(1 + f_{restore})d + \left(1 - f_{hyb}\right)C_{\mathrm{DES}}\Delta \tag{2.79}$$

Subsequently, an academic discourse on the individual variables within the equation is presented:

(1) **The Filtering Scale Δ**

The definition of the filtering scale is articulated in Eq. (2.80). To expound upon the connotations of this definition, we shall examine the case of a planar channel with a dimensional height of 2H. Figure 2.11 delineates two potential manifestations of grid scale variation. The abscissa axis corresponds to positions that extend from the wall to the center of the channel in the direction perpendicular to the wall. The ordinate axis represents the filtering scale as calculated by Eq. (2.80), which is rendered dimensionless with respect to the maximum local grid scale. The first type, as depicted by the solid line in the figure, represents the case when the grid scale perpendicular to the wall is small, that is, $\Delta_{wn} < C_w d$. According to Eq. (2.80), the value of the filtering scale is divided into three regions. When $d < \Delta_{\max}$, the filtering scale is constant, $\Delta = C_w \Delta_{\max}$. Once $d > \Delta_{\max}$, the filtering scale increases linearly with the distance to the wall d, that is, $\Delta = C_w d$. It continues until the filtering scale increases to the local maximum grid scale $\Delta_{\max}$, after which it remains constant. The alternative type, as depicted by the dashed line in Fig. 2.11, occurs when there is a rapid increase in the grid cell dimension perpendicular to the wall. In this case, the filtering scale quickly transitions from a constant value $\Delta = C_w \Delta_{\max}$ to a phase of linear increase. Due to the substantial growth rate of Δ_{wn}, reaching a state where $\Delta_{wn} > C_w d$, according to Eq. (2.80), the filtering scale $\Delta = \Delta_{wn}$, and it remains constant once it reaches $\Delta_{\max}$.

$$\Delta = \min\{\max[C_w d, C_w \Delta_{\max}, \Delta_{wn}], \Delta_{\max}\} \tag{2.80}$$

where

$\Delta_{\max}$ the local maximum grid scale is defined consistently with the methods used in DES and DDES;
Δ_{wn} the grid scale in the direction perpendicular to the wall;
C_w empirical constant, with a value of 0.15 based on WMLES method.

(2) **The Definition of the Hybrid Function** f_{hyb}

Hybrid function f_{hyb} encompasses two components: DDES and WMLES, namely:

$$\begin{aligned} f_{hyb} &= \max\left\{(1 - f_d), f_{step}\right\} \\ f_{step} &= \min\left\{2e^{-9\alpha^2}, 1.0\right\}, \quad \alpha = 0.25 - \frac{d}{\Delta_{\max}} \end{aligned} \tag{2.81}$$

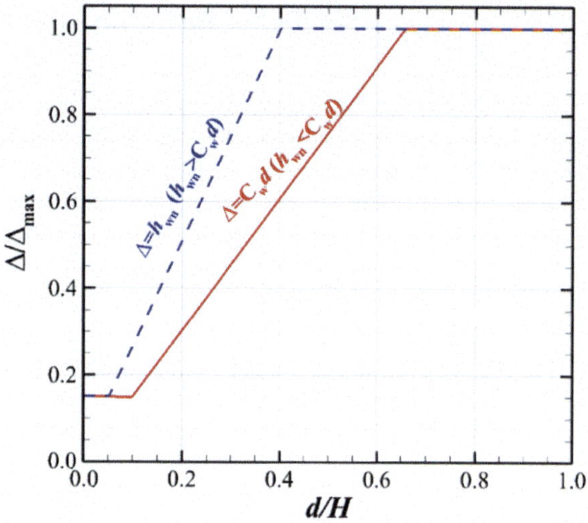

Fig. 2.11 The relationship between the filtering scale and the distance to the wall (with a channel height of $2H$) [24]

In which f_d is the transition function defined in DDES, see Eq. (2.81); f_{step} is the concept constructed in WMLES, enabling a rapid transition from a purely RANS method to a LES method within the range $0.5\Delta_{\max} < d < \Delta_{\max}$ from the wall.

(3) **The Positive Function** $f_{restore}$

To avert the occurrence of the modeled stress depletion (MSD) issue in the vicinity of the interface between RANS and LES regions, a positive function $f_{restore}$ has been incorporated:

$$f_{restore} = \max\{(f_{hill} - 1), 0\} f_{amp} \tag{2.82}$$

$$\begin{aligned} f_{hill} &= \begin{cases} 2e^{-11.09\alpha^2} & \text{if } \alpha \geq 0 \\ 2e^{-9.0\alpha^2} & \text{if } \alpha < 0 \end{cases} \\ f_{amp} &= 1 - \max\{f_t, f_l\} \\ f_t &= \tanh\left[\left(3.55^2 r_{dt}\right)^3\right], \quad f_l = \tanh\left[\left(1.63^2 r_{dl}\right)^{10}\right] \end{aligned} \tag{2.83}$$

$$r_{dl} = \frac{1}{\kappa^2 d^2} \cdot \frac{\mu_l/\rho}{\max\left(\sqrt{L_{ij}L_{ij}}, 10^{-10}\right)}; \quad r_{dt} = \frac{1}{\kappa^2 d^2} \cdot \frac{\hat{\upsilon}_T}{\max\left(\sqrt{L_{ij}L_{ij}}, 10^{-10}\right)} \tag{2.84}$$

Regarding the probe functions r_{dl} and r_{dt}, they are equivalent to the decomposition of r_d in DDES into two parts: the viscous sublayer and the logarithmic region.

2.5.5 Accuracy Validation Based on Simplified Models

Once the DES series methods were introduced, their effectiveness and accuracy were verified in simple geometric models. Table 2.3 summarizes some of these studies, where the physical phenomena validated include flow separation, vortex shedding, shock/vortex interactions, etc. It has been demonstrated that the DES method is superior to URANS in terms of predicting time-averaged parameters and the fineness of vortex resolution, especially under conditions of laminar separation, as shown in Fig. 2.12. From a qualitative perspective, the DES method can capture more refined vortex structures.

The results from simple model validations have inspired researchers to apply it to more complex high Reynolds number separated flow applications, particularly in the field of aeronautics. However, in the application to real-world engineering problems, there are still issues such as numerical stability, the contradiction between numerical accuracy and computational cost, and the grid resolution in different flow regions.

Table 2.3 Typical verification cases for the accuracy of DES methods

References	*Re*	Method	Model	Physical phenomena
Kumar [34]	1.0×10^4	SA-DES	Periodic hill	Separation, reattachment
Chen [35]	1.4×10^5	SA-DES	Cylinder	Bottom vortex separation
Deng [36]	1.4×10^5	SA-DES	Cylinder	Vortex shedding
Breuer [37]	2.0×10^4	SA-DES	Flat plate	Large separation; Kelvin–Helmholtz (K-H) instability
Jee [38]	3.9×10^3	v^2-f-DES	Cylinder	Anisotropic turbulence within the boundary layer
Zhao [39]	1.66×10^5	SST-DES; SST-DDES	Tandem cylinders	Separation flow/ boundary layer interference
Xu [40]	3.9×10^3	SST-DES; SA-DES	Cylinder	Large separation
Guseva [41]	8.51×10^5	DDES; IDDES	Blunt cylinder	Shock-vortex interaction
Nishino [42]	4.0×10^4	SA-DES	Cylinder in close proximity to a moving wall	Separation flow/ boundary layer interference
Niels [43]	$10 \sim 1.0 \times 10^6$	$k-\omega$ SST-DES coupled, $\gamma-\widetilde{Re}_\theta$ transition model	Cylinder	Transition; laminar separation; turbulent separation

(a) Computational results based on the URANS simulations employing the SST model

(b) Computational results based on the DES simulations employing the SST model

Fig. 2.12 Cylindrical flow around the vortex invariant surface [43] ($Re_D = 5 \times 10^4$)

These issues will be discussed in detail in next chapters in relation to the compressor flow field prediction problem studied in this book.

2.6 Factors Affecting the Fidelity of DES-Series Methods

DES series methods are fundamentally different from URANS methods although they are similar in the form of solving equations. URANS is modeled for turbulence at all scales, but DES-series methods are directly resolved for turbulence larger than the filtered scale in the large separation region. As a result, DES-series methods have different needs than URANS methods for the numerical solution of the N-S equations. In the following, we discuss the needs in the high-fidelity numerical solution process in terms of the factors affecting the precision and accuracy of DES-series methods, apart from gray area issues.

2.6.1 Grid Resolution

In DES-type methods, the partition of RANS and LES largely depends on the local grid scale, and in addition, in the LES solution region, the grid scale directly determines the filtering scale, so the distribution of the grid and the size of the scale (i.e., grid resolution) have a great impact on DES-type numerical methods. Specifically:

(1) **The Sparseness of the Grid Affects the Degree of Resolution of the Vortex by DES-Like Methods**

As shown in Fig. 2.13 for a cylindrical/airfoil interference problem [44], the results of DES calculations using sparse and dense meshes, respectively, clearly show that the encrypted mesh results (Fig. 2.13 right) are more reasonable for modeling the vortex shedding process. Morton et al. [45] used meshes ranging from 1.2 to 10.7 million for DES simulations of the delta wing vortex fragmentation problem, and the results of different meshes for a certain transient are shown in Fig. 2.14, which show that the denser the mesh is, the more detailed the vortex is captured, and the analyzed vortex level is more enriched. The same conclusion can be found in the studies of cylindrical flow problems [32] and double-cylindrical problems [46].

(2) **The Sparseness of the Mesh Also Has a Great Influence on the Quantitative Prediction of the Flow Field**

As shown in Fig. 2.15 for the quantitative comparison results of the above delta-wing problems, the peak dimensionless turbulent kinetic energy is getting closer and closer to the experimental result (0.5) as the mesh amount increases.

In order to improve the fidelity of the computation, many researchers have carried out grid-independent studies for different problems, and most of them have borrowed the URANS method of verifying the irrelevance, i.e., the time-averaged parameters are basically unchanged with the increase of the mesh volume. Roughly, the global time-averaged parameters are compared [45, 46], and there are also comparisons based on the local details of the time-averaged results [46]. However, time-averaging erases a lot of non-constant details, and although the time-averaged results are the same, the results are still different for different meshes from a non-constant point of

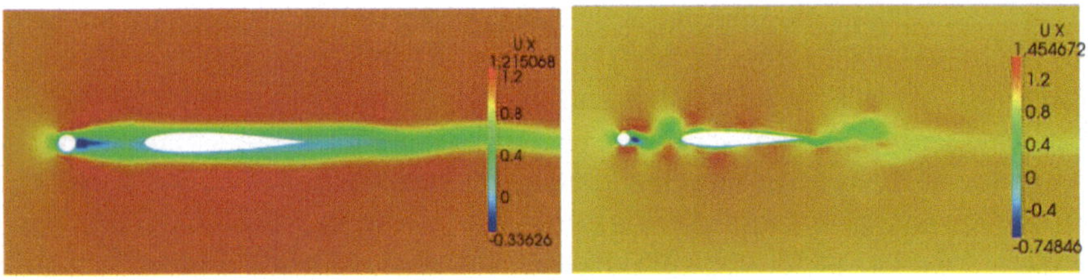

Fig. 2.13 Transient dimensionless streamwise velocity cloud (left: sparse grid; right: encrypted grid) [44]

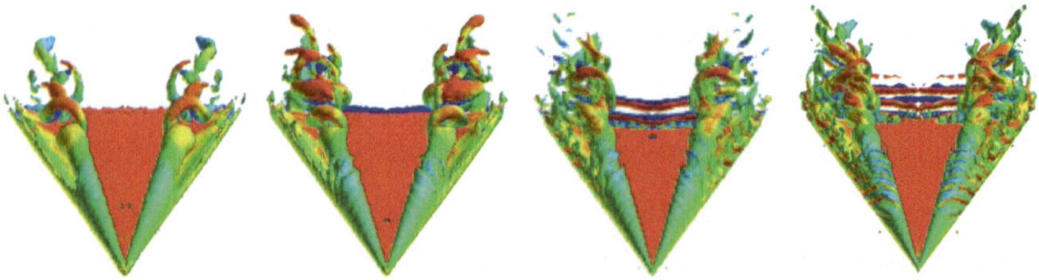

(a) 1.2 million (b) 2.7 million (c) 6.7 million (d) 10.7 million

Fig. 2.14 Delta wing transient vorticity isosurface map [45]

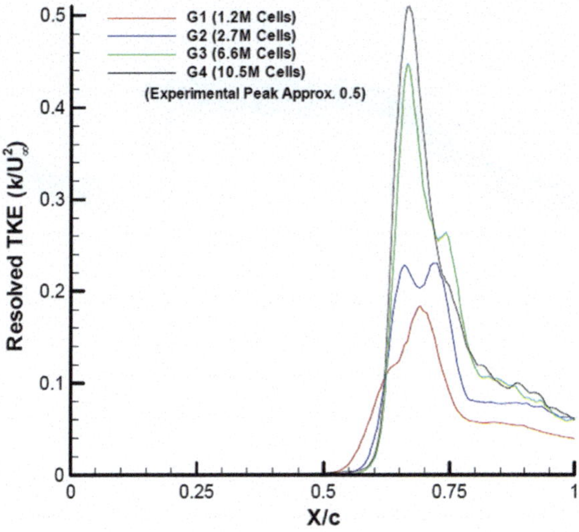

Fig. 2.15 Dimensionless turbulent kinetic energy distribution along the delta wing centerline [45] (experimental peak ~ 0.5)

view [46]. Therefore it is very difficult to find a suitable mesh resolution as a non-stationary and very mesh sensitive DES method. Travin et al. [47] serial columns indicated that it is difficult to find the convergence of the grid, which becomes a major factor affecting the accuracy of the DES-series methods.

2.6.2 *Spatio-temporal Discretization Accuracy*

Another important influencing factor is the numerical format. In order to realize the numerical solution of the control equations, it is necessary to represent the differential terms in time and space as discrete algebraic forms. The DES numerical method requires that the numerical format should be as rigorous as that of Large Eddy Simulation (LES) because most regions with large eddy separation structures are solved by LES, otherwise, small-scale eddy structures in large-scale separation regions cannot be finely captured. However, the increase in format accuracy not only increases the computational effort, but also adversely affects the stability of the computation. Therefore, it is very important to choose a suitable numerical format.

In DES-type numerical methods, both windward and central formats are used, but in terms of the characteristics of the format itself, the windward format is more stable for the problem of compressible flow field, especially when there are excitations in the flow field. In the numerical simulations of the DDES method for transonic foliation by Li [48] and Su [49], the interference between excitation and vortex is not well captured in the numerical simulations of the former using the 2nd order windward format, while the latter using the 5th order windward format captures the excitation as well as the interference between excitation and the advected surface layer very

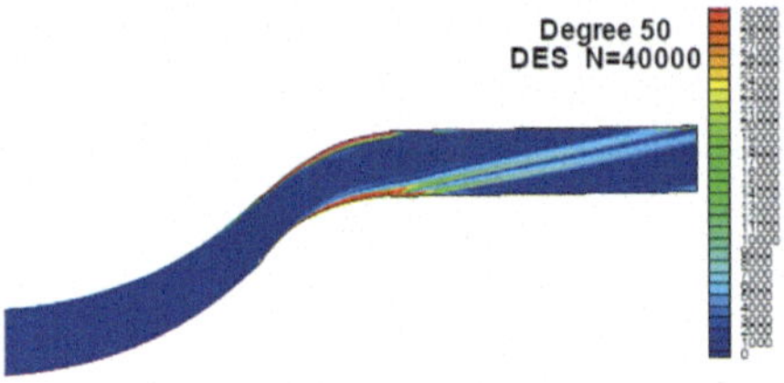

(a) Second-order center difference

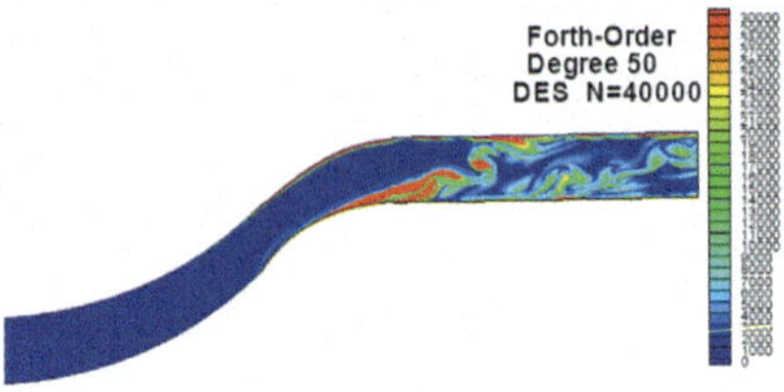

(b) Fourth-order center difference

Fig. 2.16 Transient vortex cloud for 2D vane-grid flow field of a pressurized gas engine based on DES method [20]

well. The differences in the numerical prediction results due to the format should not be ignored.

Li-Min Gao's group at Northwestern Polytechnical University [20] compares the numerical results of the second-order center difference and fourth-order center difference formats using a two-dimensional compressor vane grid as an example. As shown in Fig. 2.16, for the numerical simulation of the DES, under the same mesh size, the lower-order format (the second-order center difference) fails to distinguish the rich vortex structure due to the large numerical viscosity, which leads to the result that there is no difference between the result of the high fidelity DES and the result of the RANS, whereas the higher-order format (the fourth-order center difference) can distinguish the richer vortex structure.

For the high subsonic, strong inverse pressure gradient, and non-constant flow field studied in this book, the development of high fidelity numerical methods requires the following in terms of time discretization and space discretization.

1. **Time Discretization**

Due to the unsteady effects in the flow field, a reasonable unsteady time-stepping method is crucial for high-fidelity and efficient solutions. Here, "reasonable" means balancing discretization accuracy with computational resource consumption and the accuracy requirements of the numerical method for the discretization scheme. Time discretization schemes are divided into explicit and implicit methods. Explicit schemes, constrained by stability requirements, have very small physical time steps, requiring a long computational effort to converge. Although implicit schemes can avoid the issue of small physical time steps, each physical time step involves approximate matrix solutions, which cannot guarantee time accuracy and require substantial computational effort. This problem is especially pronounced in high-fidelity algorithms with relatively large grid sizes. Therefore, when choosing a time discretization scheme, it is necessary to balance discretization accuracy and computational efficiency.

2. **Spatial Discretization**

(1) Low dissipation: high-fidelity methods reduce the turbulent viscosity coefficient by reducing the dissipation term in the turbulence model, which requires that

the spatial discretization needs to reduce the numerical viscosity (less than the turbulent viscosity) by improving the numerical format accuracy, otherwise the flow field dominated by the numerical viscosity will not be able to correctly predict the physical viscosity in the flow field.

(2) Weak dispersion: hypersonic flow fields usually have a surge problem, and the unphysical oscillations due to dispersion can be overcome if the physical viscosity of the flow field itself acts strongly, but for high Reynolds number problems, the physical viscosity itself is weak, which puts more challenging requirements on the numerical scheme.

(3) Stability: the numerical stability of the high-precision numerical format is poor, especially for high-speed, strong reverse pressure gradient flow field, in the vicinity of the attached surface layer and the stagnation point, the great velocity gradient is more difficult to ensure the stability of the solution.

In summary, the spatial discrete format needs to minimize dissipation while ensuring numerical stability, and the dispersion of the format also needs to be considered in the strong subsonic problem.

References

1. Gao Limin, Li Kaitai, Liu Bo, et al. General form of integral conservation type N-S equation and its application in numerical simulation. Computational Physics, 2008 (02): 50–56 (in Chinese).
2. Liu Wei, Zhang Theory, Wang Yongxian, Deng Xiaogang. Computing Basics of Parallel Programming for Aerodynamics. Beijing: National Defense Industry Press, 2013.
3. Jameson A. Time dependent calculation using multigrid, with applications to unsteady flows past airfoils and wings. Honolulu: 10th Computational Fluid Dynamics Conference: AIAA, 1991.
4. Jameson A, Schmidt W, Turkel E. Equations by Finite Volume Methods Using Runge-Kutta Time Stepping Scheme. Palo Alto: 14th Fluid and Plasma Dynamics Conference, 1981.
5. Thomas H P. Artificial Dissipation Models for the Euler Equations. AIAA Journal, 1986, 24 (12): 1931–1940.
6. Swanson R C, Turkel E. Artificial Dissipation and Central Difference Schemes for the Euler and Navier-Stokes Equations. Honolulu: 8th Computational Fluid Dynamics Conference, 1987.
7. Roe P L. A survey of upwind differencing techniques. Williamsburg: 11th International Conference on Numerical Methods in Fluid Dynamics, 1988.
8. Toro, Eleuterio F. Riemann solvers and numerical methods for fluid dynamics. New York: Springer, 1999: 54–58.
9. Zha G, Shen Y, Wang B. An improved low diffusion E-CUSP upwind scheme. Computers & Fluids, 2011, 48 (1): 214–220.
10. Van Leer B. Towards the ultimate conservative difference scheme. V. A second-order sequel to Godunov's method. Journal of Computational Physics, 1997, 135 (2): 229–248.
11. Van Albada G D, Van Leer B, Roberts Jr W W. A comparative study of computational methods in cosmic gas dynamics. Astronomy and Astrophysics, 1982, 108: 76–84.
12. Harten A. High resolution schemes for hyperbolic conservation laws. Journal of Computational Physics, 1983, 49 (3): 357–393.
13. Liu X D, Osher S, Chan T. Weighted essentially non-oscillatory schemes. Journal of Computational Physics, 1994, 115 (1): 200–212.

14. Gageik M, Klioutchnikov I, Olivier H. Comprehensive mesh study for a Direct Numerical Simulation of the transonic flow at Rec, = 500,000 around a NACA 0012 airfoil. Computers & Fluids, 2015, 122: 153–164.
15. Im H S, Zha G C. Flutter prediction of a transonic fan with traveling wave using fully coupled fluid/structure interaction. San Antonio: ASME Turbo Expo 2013, 2013.
16. Li Zhen, Zhou Xun, Wang Songtao, et al. Large eddy simulation of internal flow in a plane cascade based on high-precision WENO format. Journal of Engineering Thermophysics, 2015 (4): 756–760.
17. Smit J, Annaland M V S, Kuipers J A M. Grid adaptation with WENO schemes for non-uniform grids to solve convection-dominated partial differential equations. Chemical Engineering Science, 2005, 60 (10): 2609–2619.
18. Wang R, Feng H, Spiteri R J. Observations on the fifth-order WENO method with non-uniform meshes. Applied Mathematics & Computation, 2008, 196 (1): 433–447.
19. Shen Y, Zha G. A robust seventh-order WENO scheme and ITS applications. Reno: 46th AIAA Aerospace Sciences Meeting and Exhibit, 2008.
20. Chen Xuan. Research on unsteady numerical simulation methods of impeller machinery. Xi'an: Northwestern Polytechnical University, 2014: 41–47.
21. Spalart P R, Allaras S R. A one-equation turbulence model for aerodynamic flows. Reno: 30th Aerospace Sciences Meeting and Exhibit, 1992.
22. Spalart P R. Comments on the feasibility of LES for wings, and on a hybrid RANS/LES approach//Proceedings of the First AFOSR International Conference on DNS/LES. Greyden Press, 1997: 137–147.
23. Menter F, Kuntz M, Bender R. A scale-adaptive simulation model for turbulent flow predictions//41st AIAA Aerospace Sciences Meeting and Exhibit. Reno, Nevada: AIAA, 2003.
24. Bogey C, Marsden O, Bailly C. Large-eddy simulation of the flow and acoustic fields of a Reynolds number 10 5subsonic jet with tripped exit boundary layers. Phys. Fluids, 2011, 23: 035104.
25. Lin S, Li R, Gao L, G N. Accuracy investigations of dynamic characteristic predictions of tip leakage flow using detached eddy simulation. Aerospace, 2024, 11 (1): 29.
26. Spalart P R, Deck S, Shur M L, et al. A new version of detached-eddy simulation, resistant to ambiguous grid densities. Theoretical and Computational Fluid Dynamics, 2006, 20 (3): 181–195.
27. Shur M L, Spalart P R, Strelets M K, et al. An enhanced version of DES with rapid transition from RANS to LES in separated flows. Flow Turbulence and Combustion, 2015, 95 (4): 709–737.
28. Spalart P R. Detached-Eddy Simulation. Annual Review of Fluid Mechanics, 2009, 41 (1): 181–202.
29. Chauvet N, Deck S, Jacquin L. Zonal detached eddy simulation of a controlled propulsive jet. AIAA Journal, 2007, 45 (10): 2458–2473.
30. Deck, S. Zonal-detached eddy simulation of the flow around a high-lift configuration. AIAA Journal, 2005, 43 (11), 2372–2384.
31. Deck S. Recent improvements in the zonal detached eddy simulation (ZDES) formulation. Theoretical and Computational Fluid Dynamics, 2012, 26 (6): 523–550.
32. Nikitin N V, Nicoud F, Wasistho B, et al. An approach to wall modeling in large-eddy simulations. Physics of fluids, 2000, 12 (7): 1629–1632.
33. Shur M L, Spalart P R, Strelets M K, et al. A hybrid RANS-LES approach with delayed-DES and wall-modelled LES capabilities. International Journal of Heat & Fluid Flow, 2008, 29 (6): 1638–1649.
34. Kumar G, Lakshmanan S K, Gopalan H, et al. Comparative study of hybrid RANS-LES models for separated flows. Rhodes: International Conference of Numerical Analysis and Applied Mathematics, 2015.
35. Chen Jiangtao, Zhang Peihong, Zhou Naichun, etc. Application of DES method based on SA turbulence model. Journal of Beihang University, 2012 (07): 905–909.

36. Deng Feng, Wu Yizhao, Liu Xueqiang. Numerical simulation of vortex motion in separated flow using DES. Computational Physics, 2008 (06): 683–688.
37. Breuer M, Jovičić N, Mazaev K. Comparison of DES, RANS and LES for the separated flow around a flat plate at high incidence. International Journal for Numerical Methods in Fluids, 2003, 41 (4): 357–388.
38. Jee S K, Shariff K. Detached-eddy simulation based on the v2-f model. International Journal of Heat and Fluid Flow, 2014, 46: 84–101.
39. Zhao W, Wan D. Detached-eddy simulation of flow past tandem cylinders. Applied Mathematics and Mechanics, 2016, 37 (12): 1272–1281.
40. Xu C, Chen L, Lu X. Large-eddy and detached-eddy simulations of the separated flow around a circular cylinder. Journal of Hydrodynamics, 2007, 19 (5): 559–563.
41. Guseva E K, Garbaruk A V, Strelets M K. Assessment of delayed DES and improved delayed DES combined with a shear-layer-adapted subgrid length-scale in separated flows. Flow, Turbulence and Combustion, 2017, 98 (2): 481–502.
42. Nishino T, Roberts G T, Zhang X L, et al. Unsteady RANS and detached-eddy simulations of flow around a circular cylinder in ground effect. Journal of Fluids and Structures, 2008, 24 (1): 18–33.
43. Niels N. Sørensen, A. Bechmann, F. Zahle. 3D CFD computations of transitional flows using DES and a correlation based transition model. Wind Energy, 2011, 14 (1): 77–90.
44. Samion S R L, Shaharuddin N H, Ali M S M. Grid convergence study for detached-eddy simulation of flow over rod-airfoil configuration using OpenFOAM. Bangkok: 4th International Conference on Mechanical and Aeronautical Engineering, 2018.
45. Morton S. High Reynolds number DES simulations of vortex breakdown over a 70 degree delta wing. Orlando: 21st AIAA Applied Aerodynamics Conference, 2003.
46. Lei H, Zhou D, Bao Y, Li Y, Han Z. Three-dimensional improved delayed detached eddy simulation of a two-bladed vertical axis wind turbine. Energy conversion and management, 2017, 133: 235–248.
47. Travin A, Shur M, Strelets M, et al. Detached-eddy simulations past a circular cylinder. Flow, Turbulence and Combustion, 2000, 63 (1): 293–313.
48. Li Xuesong, Gu Chunwei. Simulation study on separation vortex of high-pressure stage stator blades of compressors with large separation. Journal of Engineering Thermophysics, 2009, 30 (01): 31–34.
49. Bian Xiutao, Lin Dun, Su Xinrong, Yuan Xin. Research on the interference mechanism of shock waves and boundary layers in transonic guide vanes based on DDES. Propulsion Technology, 2017, 38 (10): 2254–2261.
50. Spalart P R. Detached eddy simulation. The Annual Review of Fluid Mechanics, 2009, 41: 181–202.

Chapter 3
High-Fidelity Numerical Simulation Method of Flow Field in Compressor Based on DES

3.1 Introduction

DES methods have good application results in simple physical models such as cylinders, tandem cylinders, and periodic mountains. However, when applied to numerical simulations of compressor flow fields, the unsteady flow fields have large spatio-temporal spans and pseudo-sequential spatial structures. Strong coupling, strong turbulence anisotropy and strong adverse pressure gradient, coupled with geometrically complex geometry with multiple walls and large curvature, are extremely detrimental to both the stability of the calculation and the accuracy of the results. Challenge. This chapter will focus on the problems that exist when DES methods are applied to compressor flow field simulation. The grid resolution is difficult to determine, the numerical stability is poor, and the calculation amount and calculation accuracy are inconsistent, and solutions will be given one by one.

The application of DES methods in the flow field inside the compressor started from Li Xuesong et al. of Tsinghua University [1, 2]. The DES method based on the S-A model was first introduced in 2007. In the numerical study of the flow field in the compressor cascade, comparison with the experimental results shows that DES is closer to the experimental results than RANS in terms of macroscopic time-averaged performance. The research on compressor cascades by Gao Limin's research group at Northwestern Polytechnical University [3] also confirmed the advantages of DES methods in capturing the fine flow field structure in the separation zone. Subsequently, Liu [4] and Hou [5] of Beijing University of Aeronautics and Astronautics used the DDES method based on the S-A model to numerically simulate the compressor cascade, and also confirmed from the time-averaged macro parameters Accuracy of DDES method for simulating three-dimensional angular anisotropic turbulence. In 2008, Gu [6], Li [7] and others made attempts to apply DES methods to compressor rotors and single-stage compressors. However, due to the lack of experimental data comparison, its accuracy cannot be determined. Know. In order to verify the accuracy of the application of DES methods in the flow field in subsonic compressors, from

R. Li et al., *Detached-Eddy Simulation Series Methods for Compressor Internal Flow*,
https://doi.org/10.1007/978-981-96-0255-1_3

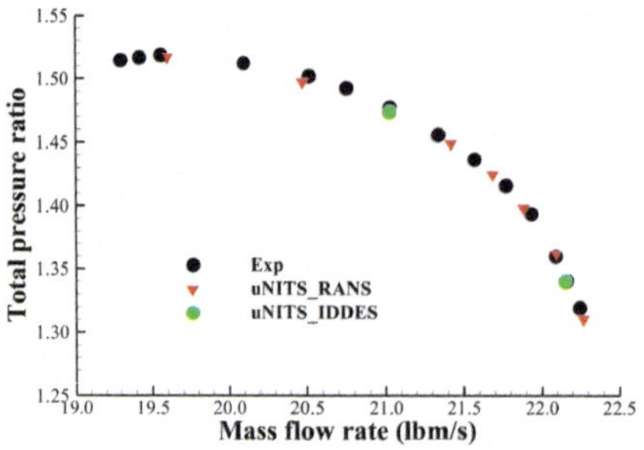

(a) Total pressure ratio

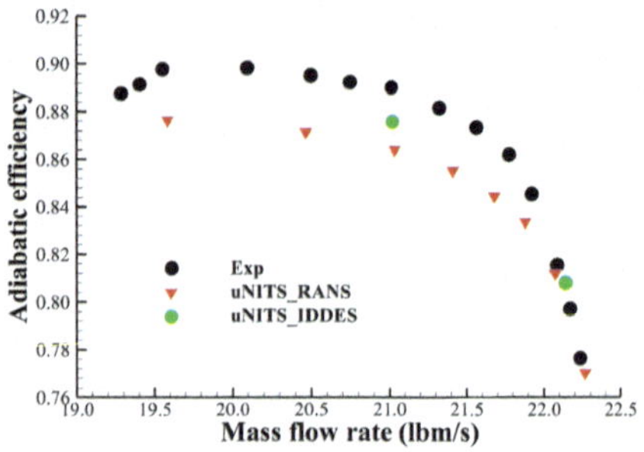

(b) Adiabatic efficiency

Fig. 3.1 Performance curve of a compressor [12]

2011 to 2016, Zha Gecheng's research group conducted experiments on DES and DDES methods in transonic single rotors [8, 9] and single-stage transonic compressor [10, 11] was checked, and it was found by comparing with the URANS method Under near-stall conditions, the DES method is more accurate in predicting macroscopic performance, and is more precise in capturing secondary flows such as gap leakage flow and corner separation. In China, the numerical simulation results of a certain transonic compressor rotor using the IDDES method by Fu Song's research group at Tsinghua University [12, 13] also proved the effectiveness of the IDDES method in predicting efficiency. The degree of accuracy is shown in Fig. 3.1.

Yamada [13] and Kusano [14] of Kyushu University in Japan compared experiments with macroscopic performance and time-averaged flow field respectively, proving the accuracy of DES application in low-speed axial flow compressors. In 2015, for the first time, the numerical simulation of the 7-stage subsonic compressor was carried out in the full ring and non-design working conditions. The calculation amount reached the peak of the application of DES method in the internal flow field. The numerical simulation and experimental comparison of the static pressure ratio of each stage were as follows As shown in Fig. 3.2, it can be seen that from the perspective of time-averaged macro parameters, the prediction accuracy of DES for the front stage is higher, but the accuracy for the later stage is lower than RANS.

For high-fidelity DES methods, the advantage lies in the capture of flow details. Verification of accuracy requires not only macroscopic experimental parameters, but also experimental results that can reflect more flow details. Detailed test data can provide a richer and more reliable basis for verifying the accuracy of high-fidelity numerical simulations. Ma et al. from the University of Lyon in France [16] conducted detailed steady and unsteady flow field measurements on a low-speed compressor cascade ($Ma = 0.12$). Since then, a large number of DES types have been developed that are very specific to this cascade. Numerical research was carried out [4, 17]. The study found that although the DES method is more accurate in loss prediction than RANS in some areas (Fig. 3.3), and also successfully captures the unsteady "double peak phenomenon", it is separated in the corner area close to the end area, as shown in Fig. 3.4c, d, there are still obvious differences with the experimental

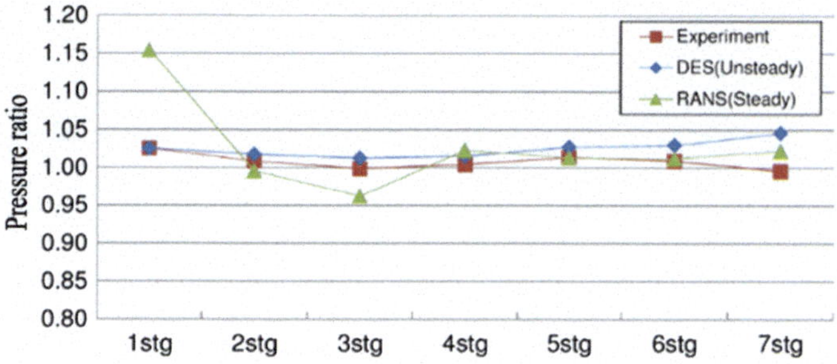

Fig. 3.2 Static pressure ratio of each stage under non-design conditions (7-stage compressor) [15]

results. It should be pointed out that these studies are all focused on the flow field of low-speed incompressible compressors. However, there is a lack of detailed flow field experimental results for high-speed compressors for verification of numerical results. The convergence of the flow field with strong adverse pressure gradient is poor. In addition, the Reynolds number is relatively high. Even if DES methods are used, the calculation amount is still large, and there are few studies on the flow field of high subsonic compressors.

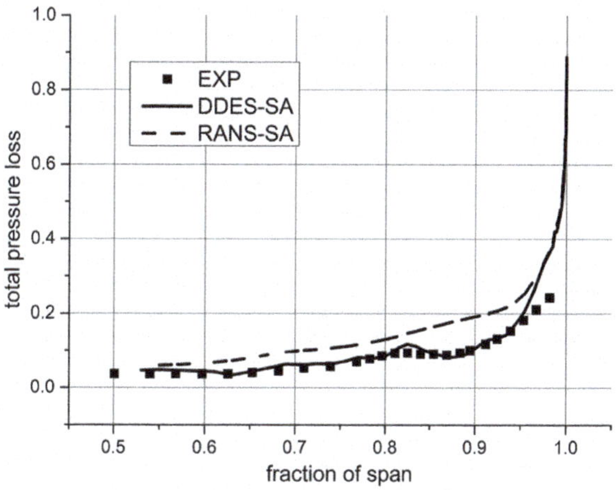

Fig. 3.3 Distribution of circumferential average total pressure loss along the span (50% chord length position) [18]

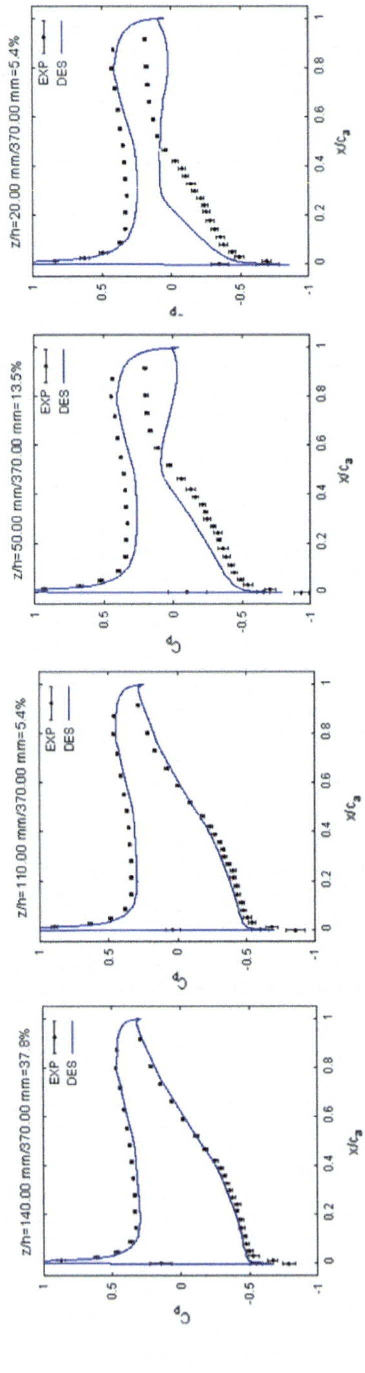

(a) 37.8%span (b) 29.7% span (c) 13.5% span (d) 5.4% span

Fig. 3.4 Comparison of DES results and experimental results of wall static pressure coefficients of different blade heights of compressor cascades [17]

3.2 Problems in Applying DES Method to Compressor Flow Field Simulation

(1) **The Rationality of Grid Resolution Is Difficult to Judge**

It can be seen from the construction principle of the DES class method that the DES class method has a strong dependence on the grid scale. Many studies have shown that the grid resolution has a great impact on the time-averaged results and the resolution of eddies. From the perspective of a single grid scale, a grid that is too dense may cause the switch from RANS to LES near the wall to be turned on in advance, leading to grid-induced separation and non-physical detachment; while a grid that is too sparse may cause numerical stickiness. The sexual contamination is too large to identify fine vortex structures, so the meshing of DES-type methods relies heavily on the user's experience.

(2) **Numerical Stability Is Difficult to Guarantee**

The flow field of a high-load compressor is characterized by strong reverse pressure and high speed. The stability of numerical calculations has become a key issue in the development of high-fidelity numerical methods for compressors. For high-precision DES methods, due to the reduction of numerical viscosity and turbulence dissipation, the numerical stability problem becomes more prominent and it is difficult to obtain a converged flow field.

(3) **Huge Consumption of Computing Resources**

Aviation compressors have the characteristics of relatively high Reynolds number and complex geometry. As the requirements for fidelity increase, more stringent requirements are placed on the spatial scale. When the spatial scale of the grid cells decreases, the physical time step, that is, the time scale, also needs to decrease accordingly to prevent smoothing out the captured high-frequency pulsations. As a result, the amount of calculation in space and time scales has increased significantly, becoming another bottleneck problem that urgently needs to be solved in high-fidelity numerical simulation.

3.3 Evaluation Method of Grid Resolution of Flow Field in Compressor

3.3.1 Grid Resolution Evaluation Criteria

Criteria will be selected considering high performance compressor design. Three grid-independence criteria are given below according to different needs (content published in Li et al. [19]):

- *Criterion 1*: Grid resolution has no impact on the time-averaged flow field. When the grid scale is reduced proportionally, the tolerance for changes in time-averaged flow field parameters (expressed as Res_{space}) does not exceed e_s. Res_{space} is defined as: $Res_{\text{space}} = |(\overline{g_1} - \overline{g_2})/\overline{g_2}|$, where $\overline{g_1}$ and $\overline{g_2}$ respectively represent the time-averaged physical quantities of the flow field before and after grid refinement. The determination of e_s depends on the accuracy requirements, usually 1 ~ 5%. A quantitative description of the grid scale is given below.
- *Guideline 2*: Grid resolution has no effect on dominant pulsations. On the basis of Criterion 1, when the grid scale is reduced proportionally, the tolerance of the main pulsation frequencies in the captured flow field (denoted as $(Res_{\text{time}})_{maj}$) does not exceed e_t. "Main pulsation" refers to the frequency contained in the modal flow field in which the pulsation energy accounts for a large proportion after POD decomposition. The definitions of $(Res_{\text{time}})_{maj}$ are given below.
- *Guideline 3*: Grid resolution has no effect on secondary pulsations. Based on Criterion 2, when the grid scale is reduced proportionally, the tolerance of the secondary pulsation frequency in the captured flow field (expressed as $(Res_{\text{time}})_{\text{min}}$) does not exceed e_t. Similarly, the "secondary pulsation frequency" here refers to the frequency contained in the modal flow field with a smaller energy share after POD decomposition.

The selection of criteria for grid resolution needs to be based on actual engineering application requirements. If you need to obtain a more accurate time-averaged flow field, then "Criterion 1" will suffice. If you need to capture the main pulsations of the flow field, then "Criterion 2" is more appropriate; if you need to capture the details of the flow field finely, then you need to select "Criterion 3". Usually, "Guideline 2" will suffice.

(1) **Selection of Indicator Parameters**

In the grid resolution evaluation, Res_{space} and Res_{time} are defined as index parameters. The selection of this parameter needs to be combined with the physical quantities of interest in compressor design and mechanism analysis.

In Criterion 1, the index parameters Res_{space} usually select the overall machine performance, such as pressure ratio, efficiency and stability margin, or local time-averaged aerodynamic parameters, such as static pressure distribution, total pressure loss coefficient and wall shear stress. wait. The considerations in this regard are the same as the grid-independent analysis in URANS and will not be repeated here.

Criteria 2 and 3 describe the impact of grid resolution on the dynamic flow field through Res_{time}. In the unsteady flow field, more attention is paid to the unsteady frequency, which is usually an important indicator used in stall warning signals or unsteady flow control design. Therefore, the size of the unsteady frequency is selected as the index. This unsteady frequency can be the result of POD decomposition of physical fields such as pressure field, velocity field or temperature field, and then extracting the main pulsating modes for FFT transformation. Compared with secondary pulsations, primary pulsations have higher pulsation energy and larger vortex scales, which is why we choose to use POD for analysis. Table 3.1 taking the hot aerodynamic issues in the compressor field as an example, the difference between the main pulsation frequency and the secondary pulsation frequency is given.

(2) **Definitions of Res_{time} and e_t**

Assume that the frequencies corresponding to the k-order modes predicted by low grid density and high grid simulation are f^k and f^k_{std} respectively, then the definition of Res_{time} as follows:

$$Res_{\text{time}} = \left| \frac{f^k - f^k_{\text{std}}}{f^k_{\text{std}} \cdot f^k} \right| \tag{3.1}$$

Error range e_t is defined as

$$e_t = N_{\min} \Delta t_{\text{physical}} \tag{3.2}$$

Among them, $N_{\min}$ represents the minimum number of samples describing a cycle, $\Delta t_{\text{physical}}$ represents the physical time step. For example, for a sine wave, at least 5 points are needed to describe a cycle, that is, $N_{\min} = 5$. The meaning of this definition is that as the grid gradually becomes denser, based on the calculation settings of

Table 3.1 Meaning of frequency in typical compressor aerodynamic problems

Research problem	Main frequency	Minor frequency
Rotating stall [20]	Frequencies of rotating stall cell	Frequencies related to vortex separation or shedding
Vortex induced blade vibration [21]	The frequencies of pressure fluctuation on the blade surface	Frequencies not directly related to vibration
Flow-induced noise [22]	Dynamic pressure fluctuation frequency with high power spectrum density (dB)	Dynamic pressure fluctuation frequency with low power spectrum density (dB)
Flow loss [23]	Large-scale fluctuation frequency in high loss region. Such as frequencies of tip leakage vortex in tip clearance region and frequencies of passage vortex, corner vortex in corner separation region	Pulsation frequency in low loss area
Unsteady flow control [24]	Eddy frequency induced by flow control technology	Frequencies not directly related to unsteady flow control
Rotor–stator interference [25]	Frequencies of wakes Frequencies of large eddy frequency close to tip region	Fluctuation frequencies related to the vortexes inside the wakes

the current physical step size, a new sine wave cannot be captured within a period corresponding to the frequency error.

(3) **Quantitative Description of Grid Scale**

The quantitative description of the grid scale intuitively reflects the size of the grid resolution and improves the portability of grid generation. That is, once the grid scale is determined, it can be used as a reference when generating geometrically similar grids without repeating the grid. Grid resolution assessment. This can effectively reduce the cost of grid resolution research based on high-fidelity numerical simulation research, while improving the reliability of calculation results.

To quantitatively describe the grid scale of a certain area, we define a parameter named average dimensionless grid scale, represented by $\overline{\Delta l^+}$. It represents the average value of the resolved minimum turbulence length scale when performing LES numerical simulations. The definition of the length scale Δl depends on the filtering scale in the DES class method. There are three commonly used definitions in the literature. Taking the grid k as an example, the filtering scale of the local grid can be defined as: ① The local maximum grid scale (Formula (3.3)) [26]; ② The local unit volume Cube root (Formula (3.4)) [27]; ③ Length scale based on vorticity (Formula (3.5)) [28].

$$\Delta l_k = \max(\Delta x_k, \Delta y_k, \Delta z_k) \tag{3.3}$$

$$\Delta l_k = (\Delta x_k \Delta y_k \Delta z_k)^{1/3} \tag{3.4}$$

$$\Delta l_k = \sqrt{N_{x,k}^2 \Delta y_k \Delta z_k + N_{y,k}^2 \Delta x_k \Delta z_k + N_{z,k}^2 \Delta x_k \Delta y_k}, \ \mathbf{N} = \frac{\boldsymbol{\psi}}{\|\boldsymbol{\psi}\|}, \ \boldsymbol{\psi} = \nabla \times \mathbf{W} \tag{3.5}$$

The dimensionless grid scale Δl_k^+ is defined as:

$$\Delta l_k^+ = Re_\tau \frac{\Delta l_k}{C} \tag{3.6}$$

Here C represents characteristic dimensions, usually compressor blade chord length or rotor blade height. Re_τ represents the Reynolds number based on wall shear stress. Re_τ at grid cell k is defined as:

$$Re_{\tau,k} = \frac{u_{\tau,k} \cdot C}{\mu_{l,k}/\rho_k}, \ u_{\tau,k} = \sqrt{\frac{\tau_{w,k}}{\rho_k}} \tag{3.7}$$

where $\mu_{l,k}$ and $\tau_{w,k}$ represent the laminar turbulence coefficient and the magnitude of the wall shear stress at grid cell k, respectively. Re_τ is defined as the maximum of $Re_{\tau,k}$. Considering that the grid cells close to the leading edge and trailing edge are generally refined, the maximum value is selected after removing the values within the range of 5%C near the two edges, that is:

$$k \in \{\text{grid cells of blade surface in the range of } 5\%C \sim 95\%C\} \tag{3.8}$$

Considering that the grid of the entire computational domain is non-uniformly distributed, for the convenience of discussion, the dimensionless maximum grid scale averaged based on the grid unit volume (Ω_k) in the specified computational domain is used as the index, and the symbol is expressed as $\overline{\Delta l^+}$, that is:

$$\left.\overline{\Delta l^+}\right|_{\text{region}} = \sum_{k=1}^{cellNumber} \Delta l_k^+ \cdot \Omega_k / \sum_{k=1}^{cellNumber} \Omega_k \tag{3.9}$$

3.3.2 Compressor Flow Field Zoning Strategy

In 2001, Spalart [29] provided grid guidelines for external flow field simulations using the DES method, such as aerofoils. Similarly, according to the flow characteristics of multiple walls, complex physical phenomenon like shockwave and vortex in the compressor's flow field, the computational domain can be divided into four parts: the focus region (FR), RANS-LES region (RL), inlet region (IR), and RANS region (RR). Take a three-dimensional (3D) compressor rotor blade, ROTOR37, as an example, as presented in Fig. 3.5. The characteristics of each region are listed below:

(1) **FR region**: The most critical region is colored red in Fig. 3.5, which contains significant complex flow phenomena in the region, such as separation, shock wave, and vortex interaction. In the turbomachinery flow field, complexity always appears near the blade tip, and hub corner due to tip leakage flow, corner separation, wake, and they have a significant effect on its performance, which deserves detailed analysis. The turbulence in the region is characterized

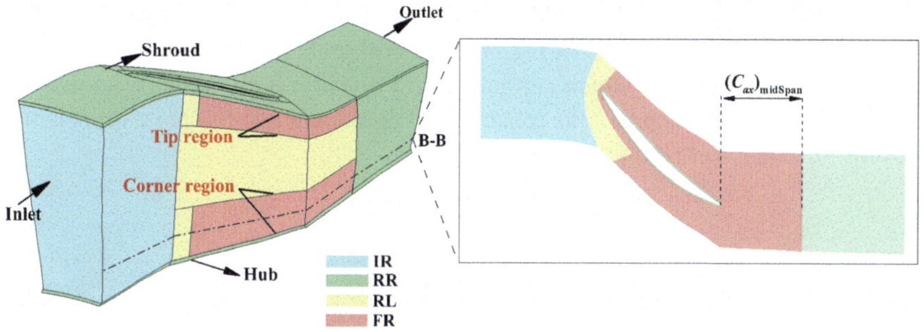

Fig. 3.5 Schematic diagram of grid partition

by anisotropy and nonequilibrium. Thus, it cannot be adequately described by the RANS or URANS method with the eddy viscosity turbulence model. The grid-scale in this region is relatively small compared with others.

(2) **RL region**: The region is located in the blade passage without the complex flow phenomena mentioned above and is colored yellow in Fig. 3.5. Controlling the airflow through the blade passage is the basic principle of turbomachinery; thus, the passage flow deserves our attention. While the flow fields in some regions are relatively simple, such as the region near the mid-span or with an accelerating pressure gradient, we define it as the RL region.

(3) **IR region**: The region contains inlet extension from the inlet to the position near the leading edge and is colored blue in Fig. 3.5. The flow field in the region is relatively uniform and stable. The grid number in this region is relatively small due to the low-pressure gradient and unresolved inlet flow unsteadiness.

(4) **RR region**: The region close to the solid walls and outlet is colored green in Fig. 3.5. The RANS method is expected to be employed in the region. The grid number in the boundary layer close to the wall is approximately 20. The region one chord away from the trailing edge is also defined as the RR region because the intensity of flow mixing has been reduced, and the large-scale vortex has been considerably dissipated.

3.3.3 Grid Resolution Evaluation Process

The grid resolution evaluation flow chart is shown in Fig. 3.6. First, it is necessary to clarify the requirements for the accuracy of numerical simulation. The three requirements correspond to the three criteria respectively. Next, partition the entire computational domain. Then, mesh refinement is performed in different areas, especially the FR and RL areas, and the mesh refinement increases in a certain proportion according to the dimensionless mesh scale $\overline{\Delta l^+}$. Then the DES method is used to numerically simulate the numerical models with different grid resolutions. For "Requirement 1", it is enough to analyze the time-averaged flow field to meet Criterion 1. For "Requirement 2" or "Requirement 3", based on the analysis of the time-averaged flow field, analyze whether the pulsating flow field satisfies Criterion 2 or 3. When it is not satisfied, continue to refine the mesh for numerical calculation and analysis.

After the resolution analysis of Fig. 3.6, the grid scale for different needs can be quantitatively determined, which can guide grid division in numerical simulations of similar geometries, improve the reliability of numerical results, and eliminate the need for repeated grid resolution, rate analysis.

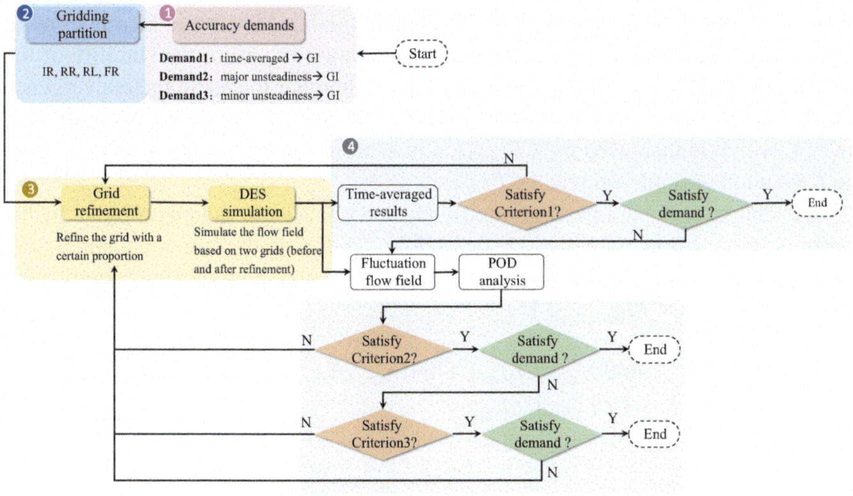

Fig. 3.6 Flow chart of grid resolution evaluation

3.4 Numerical Stability Processing Method for Strong Adverse Pressure Flow Field

3.4.1 CFL *Increasing Number Method*

The CFL number is the ratio of the distance the waveform perturbation propagates in a time step to the grid size, and its size depends on the CFL (Courant–Friedrichs–Lewy) stability condition. Stability conditions are necessary conditions for ensuring format stability in CFD and play an important role in the stability of numerical methods. From a mathematical point of view, when discrete methods are used to stably solve hyperbolic partial differential equations, the dependence domain of the difference equation must include the dependence domain of the corresponding differential equation; from a CFD perspective, the stability condition indicates that for each Computational units, the distance any information travels within a time step must be much smaller than the distance between grid cells. According to the stability conditions, in hyperbolic partial differential equations like the N-S equation, the *CFL* number is generally less than 1 and near 1, so that the propagation along the characteristic line does not deviate too far or too close, which can ensure the accuracy of numerical solutions. However, in high-precision numerical calculations, such values are difficult to converge for high-speed and strong adverse pressure gradient flow fields. Although reducing the number of *CFL* can ensure the stability of the calculation, if the number of *CFL* is too small, it will slow down the convergence speed, thereby increasing the amount of calculation.

After research, it was found that for high-speed and strong adverse pressure flow field problems, divergence is prone to occur in the initial stage of iteration. Therefore, this article proposes the *CFL* number increment method. The basic idea is to use a smaller *CFL* number at the beginning of the iteration to ensure the stability of the calculation, and then gradually increase the *CFL* number to ensure the convergence speed. See Formula (3.10) for details. In the initial stage of iteration *CFL* number is CFL_{init}, which is increased by CFL_{increase} in each iteration step Gradually increase the *CFL* number until you reach $CFL_{\max}$.

The value of each parameter is determined according to the actual problem. In this article $N_{\text{increase}} = 500$, $CFL_{\text{increase}} = 0.1$, $CFL_{\max} = 0.8$; in the numerical simulation of the transonic rotor (ROTOR37) $CFL_{\text{init}} = 0.001$. For the high subsonic compressor cascade flow field, $CFL_{\text{init}} = 0.1$.

$$CFL = \begin{cases} CFL < CFL_{\max} & CFL_{\text{init}} + \text{int}\left(n_{\text{step}}/N_{\text{increase}}\right) * CFL_{\text{increase}} \\ CFL \geq CFL_{\max} & CFL_{\max} \end{cases} \tag{3.10}$$

3.4.2 *Local Entropy Correction of Numerical Formats*

The Roe scheme has the advantages of low dissipation, high shock wave resolution and stable calculation. However, at high Mach numbers, especially when flow field discontinuities occur, non-physical solution will also occur. This phenomenon is called "Carbuncle" phenomenon [30]. That is, in places with large velocity gradients, the eigenvalues of Eq. (2.29) in Chap. 2 of the coefficient matrix in the Roe format are too small, violating the entropy condition, resulting in the inability to correctly simulate the expansion process. In actual calculations, the calculation results are as follows: near the shock wave, at the stagnation point of the blade leading edge, and in the boundary layer near the wall in the acceleration zone, the contours of physical parameters (pressure or Mach number, etc.) appear wrinkles, jagged and other non-physical phenomena, this unphysical oscillation may even cause numerical divergence.

In order to solve this problem, it is usually necessary to introduce entropy correction [31] in the flow field, which essentially suppresses non-physical oscillations by introducing numerical dissipation. To minimize the contamination of the flow field in weak gradient areas and large separation areas by the introduced numerical dissipation, the local entropy correction method is used here, that is, correction is only performed in areas with particularly small eigenvalues, as shown in the following formula:

$$\lambda_i = \begin{cases} \left(\lambda_i^2 + \xi^2\right)/2\xi & \lambda_i \leq \xi \\ \lambda_i & \lambda_i > \xi \end{cases} \quad \xi = \lambda_{\max}\xi_0 \tag{3.11}$$

In the formula:

ξ_0 Small quantity, we set it as 0.125.

3.4.3 Variable Limitation of Turbulence Model

The source term (the sum of the production term and the dissipation term) of the S-A model Eq. (2.62) is non-zero in a physical sense, that is to say, the production term $c_{b1}\rho\hat{S}\hat{\upsilon}_T$ must be greater than the dissipation term $c_{w1}f_w\rho(\hat{\upsilon}_T/d)^2$, the non-zero condition still needs to be satisfied even when the vorticity is zero. However, in the initial stage of numerical calculation, a reasonable velocity distribution may not be given near the wall, making the vorticity ψ_{ij} in Eq. (2.65) very small. Currently, the variables in the term $\hat{S}$ are generated. Close to zero, this will cause the production term to be smaller than the dissipation term. Subsequently, this phenomenon will cause the flow field pressure to be negative through the transfer of the energy equation, which will cause the calculation of the entire flow field to diverge.

In order to improve calculation stability, according to NASA recommendations [32], first limit the variables $\hat{S}$ in the production term so that its minimum value will not be less than $0.3S$. In addition, the dissipation term. The variables r in are limited to be no less than 10 to prevent the dividend in the dissipation term from being close to zero. See Formula (3.12) for details. Normally, such restrictions will not affect the convergence results, but will only bring the variables back to a reasonable value range in the initial stage of iteration.

$$
\begin{aligned}
\hat{S} &= \max\left[S + \frac{1}{Re_{ref}}\frac{\hat{\upsilon}_t}{\kappa^2 d^2} f_{v2},\ 0.3S\right] \\
r &= \min\left[\frac{1}{Re_{ref}}\frac{\hat{\upsilon}_T}{\hat{S}\kappa^2 d^2},\ 10\right]
\end{aligned}
\tag{3.12}
$$

3.5 Boundary Condition Processing Method for Internal Flow of Compressor

The governing Eq. (2.1) can correctly describe the motion laws of compressible Newtonian fluid, but for each individual physical problem, appropriate initial conditions and boundary conditions need to be given to obtain the specific flow field. The above two are collectively called definite solution conditions. The boundary conditions of the compressor include inlet boundary conditions, outlet boundary conditions, relatively no-slip wall boundaries, periodic boundary conditions, etc.

In order to facilitate unified iterative calculation of the entire flow field and improve parallel efficiency, this book uses the virtual grid method to process the boundaries of the computational domain. As shown in Fig. 3.7, by giving the values at virtual grid (ghost cell) $g1$, $g2$, the physical quantities at the boundary are indirectly made to satisfy the boundary Conditions, that is, converting given boundary conditions into solutions for virtual grid variable values. Considering the highest adopted 5th-order WENO scheme, two virtual grids can be extended outwards.

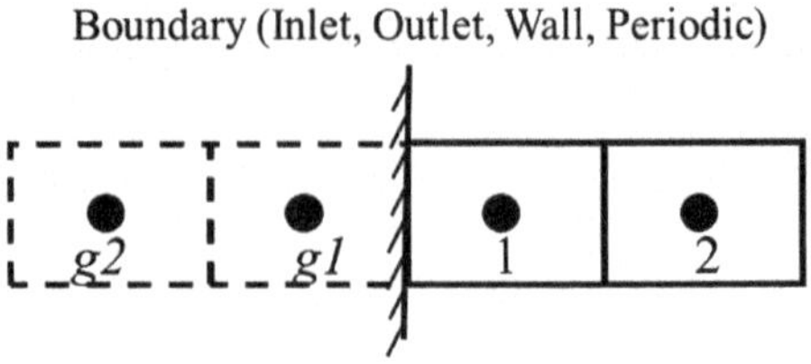

Fig. 3.7 Virtual grid diagram

3.5.1 Inlet Boundary Conditions

When the inlet airflow is subsonic, the total temperature T^*_{in}, total pressure p^*_{in} and inlet axial airflow angle θ_{axial} (x direction) are generally given. And the inlet spanwise airflow angle θ_{span} (z direction), while adopting additional conditions $\partial p/\partial l = 0$ (l means along the streamline direction), from the flow Transfer static pressure to the boundary within the field p_{in}. When using the S-A equation model, the inlet turbulent viscosity coefficient μ_t also needs to be given. Since the inlet of the compressor calculation domain has a relatively long extension, the incoming flow conditions are approximately unchanged in the flow direction. Therefore, the values of $g1$ and $g2$ virtual grid are the same as those of the grid. The values of point 1 are the same, that is: $m_{g2} = m_{g1} = m_1$, where m represents the physical quantity T^*_{in}, p^*_{in}, θ_{axial}, θ_{span}, p_{in} and μ_t. Therefore, each physical quantity on the virtual grid can be solved by solving the inlet Mach number Ma_{in} (Eq. 3.13), the inlet static temperature T_{in} (Eq. 3.14) and the inlet degree vector $\mathbf{W}_{\text{in}}$ (Eq. 3.15) is obtained.

$$Ma_{\text{in}} = \left\{ \frac{2}{\gamma - 1} \left[\left(\frac{p^*_{\text{in}}}{p_{\text{in}}} \right)^{\frac{\gamma-1}{r}} \right] \right\} \tag{3.13}$$

$$T_{\text{in}} = T^*_{\text{in}} \left(\frac{p_{\text{in}}}{p^*_{\text{in}}} \right)^{\frac{\gamma-1}{\gamma}} \tag{3.14}$$

$$|\mathbf{W}_{\text{in}}| = Ma_{\text{in}} \sqrt{kRT_{\text{in}}} \tag{3.15}$$

$$\begin{aligned} &w_x = |\mathbf{W}_{\text{in}}| \sin(\beta_{\text{span}}) \cos(\beta_{\text{axial}}); \quad w_y = |\mathbf{W}_{\text{in}}| \sin(\theta_{\text{span}}) \sin(\theta_{\text{axial}}); \\ &w_z = |\mathbf{W}_{\text{in}}| \cos(\theta_{\text{span}}) \end{aligned} \tag{3.16}$$

For the S-A turbulence model, the inlet turbulent viscosity coefficient is 5–10 times the laminar flow viscosity coefficient, that is, $\mu_t/\mu_l = 5 \sim 10$.

3.5.2 Outlet Boundary Conditions

The compressor outlet generally has subsonic boundary conditions. Only the outlet static pressure p_{out} needs to be given. Other parameters such as density, velocity, and turbulent viscosity coefficient are first-order extrapolated along the flow direction, that is:

$$\begin{aligned} p_{g1} &= 2p_{out} - p_1; \ \mathbf{W}_{g1} = \mathbf{W}_1; \ \rho_{g1} = \rho_1; \ \mu_{tg1} = \mu_{t1} \\ p_{g2} &= 2p_{out} - p_2; \ \mathbf{W}_{g2} = \mathbf{W}_1; \ \rho_{g2} = \rho_1; \ \mu_{tg2} = \mu_{t1} \end{aligned} \tag{3.17}$$

3.5.3 Relatively No-Slip Boundary Conditions

The no-slip boundary condition is an object surface boundary condition, that is, the fluid is stationary relative to the object surface. In addition, this book assumes that the physical wall is adiabatic isentropic, with $\partial p/\partial n = 0$, $\partial T/\partial n = 0$ at the wall. For this reason, the virtual grid static pressure and static temperature are given as:

$$\begin{aligned} p_{g1} &= p_1; \ T_{g1} = T_1 \\ p_{g2} &= p_2; \ T_{g2} = T_2 \end{aligned} \tag{3.18}$$

Due to the relatively slip-free boundary conditions, the velocity at the boundary is calculated slightly differently in the stator and rotor components:

Stator parts: at the boundary $w_x = 0$, $w_y = 0$, $w_z = 0$, in order to achieve the above The object surface boundary and the virtual grid physical quantity given method are:

$$\mathbf{W}_{g1} = -\mathbf{W}_1; \ \mathbf{W}_{g2} = -\mathbf{W}_2 \tag{3.19}$$

Rotor component: In the relative coordinate system, the fluid is stationary relative to the surface of the rotating object (rotor blades and hub), and the casing wall rotates counter-rotating relative to the rotor, as shown in Fig. 3.8. Given the static wall boundary on the rotor surface, the physical quantities of the casing wall in the radial, tangential, and flow direction velocities (w_{r}, w_{t}, w_{a}) in the cylindrical coordinate system must satisfy $w_r = 0$, $w_t = -\omega r$, $w_a = 0$. In the rectangular coordinate system, there is:

$$w_x = 0; \ w_y = -\omega Z; \ w_z = \omega Y. \tag{3.20}$$

Therefore, the virtual grid physical quantity given method is:

$$\begin{aligned} w_{x|g1} &= -w_{x|1}; \ w_{y|g1} = -2\omega Z - w_{y|1}; \ w_{z|g1} = 2\omega Y - w_{z|1} \\ w_{x|g2} &= -w_{x|2}; \ w_{y|g2} = -2\omega Z - w_{y|2}; \ w_{z|g2} = 2\omega Y - w_{z|2} \end{aligned} \tag{3.21}$$

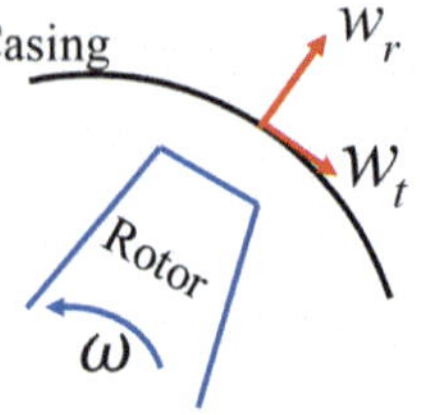

Fig. 3.8 Schematic diagram of the relative movement between the casing and the rotor

For the S-A turbulence model, at the wall $\mu_t = 0$, the value of the virtual grid is:

$$\mu_{tg1} = -\mu_{t1};\ \ \mu_{tg2} = -\mu_{t2} \tag{3.22}$$

3.5.4 *Periodic Boundary Conditions*

When the incoming flow is uniform and the compressor is operating stably, the flow field can be approximately considered as a periodic flow field with the blade channel as the period. Therefore, the simulation of the entire flow field can be reduced to the numerical simulation of a single channel, which can effectively reduce amount of calculation. In numerical simulation, periodic boundary conditions need to be given on both sides of the channel. The magnitude of each physical quantity at the two boundaries that are periods to each other is equal, that is:

$$\begin{aligned} &p_{g1} = p_1;\ T_{g1} = T_1;\ \rho_{g1} = \rho_1;\ \left|\mathbf{W}_{g1}\right| = |\mathbf{W}_1|;\ \mu_{tg1} = \mu_{t1} \\ &p_{g2} = p_2;\ T_{g2} = T_2;\ \rho_{g2} = \rho_2;\ \left|\mathbf{W}_{g2}\right| = |\mathbf{W}_2|;\ \mu_{tg2} = \mu_{t2} \end{aligned} \tag{3.23}$$

The periodic boundary given for the velocity vector is slightly different: for a planar problem, such as a compressor plane cascade, the velocity vector is the same in the Cartesian coordinate system, such as Fig. 3.9a shown in, that is, $\mathbf{W}_{g1} = \mathbf{W}_1$; for annular problems, such as compressor rotor, stator, etc., such as Fig. 3.9b, the velocity vector at the periodic boundary Equality is a condition satisfied in the cylindrical coordinate system, but in the rectangular coordinate system:

$$\begin{aligned} &\theta_1 = \arctan(Y_1/Z_1);\quad \theta_{g1} = \arctan\left(Y_{g1}/Z_{g1}\right) \\ &w_{r1} = w_{z1}\cos(\theta_1) + w_{y1}\sin(\theta_1);\quad w_{t1} = -w_{z1}\sin(\theta_1) + w_{y1}\cos(\theta_1) \\ &w_{x|g1} = w_{a1};\quad w_{y|g1} = w_{r1}\sin(\theta_{g1}) + w_{t1}\cos(\theta_{g1}); \\ &w_{z|g1} = w_{r1}\cos(\theta_{g1}) - w_{t1}\sin(\theta_{g1}) \end{aligned} \tag{3.24}$$

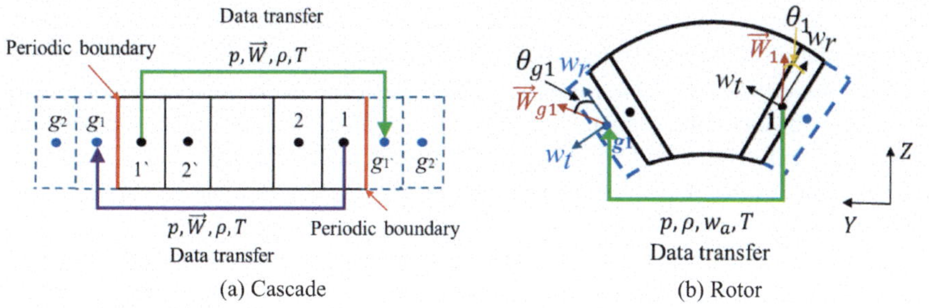

Fig. 3.9 Schematic diagram of different types of periodic boundary conditions

3.6 Parallel Method for Numerical Simulation of Flow Field in Compressor

3.6.1 Overview of Parallel Computing

Parallel computing refers to decomposing a given application problem into several independent subtasks on a parallel computer and assigning them to different processors through a certain combination. Each processor executes each task in parallel through mutual cooperation. Subtasks to finally obtain the solution to the original problem. Parallel computing is to achieve the purpose of improving the speed of solving or expanding the scale of solving problems on the premise of ensuring accurate calculation results.

For traditional serial CFD solvers, improving the solution speed involves two primary approaches. The first is enhancing the convergence rate of the algorithm itself, and the second is reducing the program's runtime by optimizing the execution method of the program code. However, for a given CFD algorithm, there are often limitations to how much the convergence speed can be improved, making parallel computing—aimed at reducing program runtime—the main method for accelerating CFD solutions.

Moreover, serial CFD solvers are constrained by the memory resources of a single processor, limiting the grid size they can handle. For instance, a 2 GB memory typically accommodates a grid size of around 100,000 cells. When numerical simulations require grids with millions or even tens of millions of cells, parallel computing becomes essential for scaling up the problem size linearly.

In the numerical simulation of internal flows in turbomachinery, serial CFD programs are generally adequate for solving the flow field within a single-stage, single-blade passage. However, for multi-stage and multi-blade passage flow fields, where the grid size is significantly larger, parallel CFD methods are necessary. Additionally, improving the accuracy of numerical simulations in turbomachinery often requires the adoption of higher-precision numerical methods. These methods, however, introduce more complex interpolation schemes or numerical calculations, thereby increasing the computational load and solution time.

To address this, parallel computing can be employed to speed up calculations by leveraging multiple processing cores to handle the increased computational demand. Therefore, achieving faster solution speeds, larger solution scales, and higher solution accuracy in the field of turbomachinery necessitates the use of parallel CFD technology.

In recent years, parallel computational fluid dynamics has developed rapidly under the strong promotion of parallel computing technology. In-depth research on the complex flow field inside the fluid machinery, especially the large-scale numerical simulation of the unsteady flow of the entire fluid machinery, requires efficient Powerful support for parallel computers and massively parallel computational fluid dynamics. Judging from the major research plans based on parallel computing released by the United States, Japan, the European Union and other countries in recent years, almost all of them list large-scale parallel computational fluid dynamics as an important support direction. Such as the National Science Foundation (NSF) [33], Department of Energy (DOE) [34], Department of Defense (DOD) [35], Aeronautics and Space Administration (NASA) [36] has supported a number of high-performance computational fluid dynamics research projects and has made progress in the fields of digital wind tunnels, turbulence mechanisms, aero-engines, and flow simulation around the entire aircraft. At present, my country's high-performance parallel computer technology is developing rapidly. During the "Eleventh Five-Year Plan" period, the Ministry of Science and Technology deployed the 863 major project "High-Performance Computer and Grid Service Environment". The National University of Defense Technology, Shuguang Company, and the 56th General Staff Institute successively launched the project in 2010. Three petaflop high-performance computer systems were developed. At the same time, major projects support a number of computing applications with a scale of around 4096 cores, including scientific computing such as high-energy physics, computational chemistry, biological information, and massive data processing, as well as railway freight, weather forecasting, water conservancy information, energy conservation and emission reduction, industrial simulation and optimization design and other engineering applications. It can be seen that high-performance parallel computing technology and its application in fluid mechanics and other aspects have attracted much attention at home and abroad.

In order to facilitate the development of parallel computing applications on different parallel computer systems, various programming models have emerged. The parallel programming model [37] is a programming interface that simplifies programmers to write parallel programs. This interface hides the communication, synchronization and other program implementation details in the parallel programming process to a certain extent. It is closely related to the structure of the parallel computer system. Related. Today's popular parallel programming models are mainly divided into two categories: shared memory model and message passing model. The most representative shared memory model is the OPENMP standard [38] originated from the ANSI Program parallelization can be achieved, and the OpenMP standard is simple and easy to use, and is widely used in fine-grained parallel CFD. The message passing model supports a multi-address space non-shared memory system. In this model, each process participating in the calculation has an exclusive local memory space, and data exchange between processes is achieved by sending and receiving messages. PVM [39] (Parallel Virtual Machine) launched by Oak Ridge National Laboratory (ORNL) in the United States in 1989 is an early parallel programming software that adopts the message passing model. Later, the message passing interface (MPI) was also released by the MPI committee in 1994 [40]. The parallel programming model based on message passing has been widely used in coarse-grained CFD parallelism.

During the parallelization process of serial algorithms or programs, the following four basic characteristics generally need to be considered: First, it is necessary to ensure that the parallel calculation results after serial CFD parallelization are consistent with the previous serial calculation results, that is, serial and parallel calculations Consistent, which is a requirement for correctness in parallel computing. Secondly, the parallelized CFD program needs to have higher performance as much as possible on the parallel computing platform it is running on, that is, the calculation can be completed at a faster speed, which is also the focus of parallel computing research. Thirdly, parallel CFD programs need to have good scalability, that is, as the number of processors increases, the size of the problem that needs to be solved is minimized while ensuring that the performance of the parallel algorithm or program remains unchanged. Finally, the parallel CFD program also needs to have good portability, that is, the parallel program can run on different parallel platforms, and its parallel performance will not fluctuate significantly as the parallel platform changes. In short, the implementation of parallel CFD requires not only the correctness and performance of the algorithm or program, but also the scalability and portability of the algorithm or program itself. This chapter focuses on introducing parallel methods based on OpenMP and MPI.

3.6.2 Parallelism Based on OpenMP

OpenMP (Open specification for multi-processing) is a thread-based parallel programming model. It is not a new language. It is an extension of existing programming languages. It supports FORTRAN, C, C++, etc., and is a shared storage architecture. Application program interface (API). Because it is easy to implement and there are many iterative calculations in three coordinate system directions in three-dimensional computational fluid dynamics. Therefore, embedding OpenMP parallel algorithms in computing programs can improve computing efficiency to a certain extent.

Hardware layers suitable for OpenMP are shared memory multi-processor systems and multi-core processors, etc. This can be achieved with the GCC[1] (GUN compiler

[1] GCC (GNU C Compiler) under Linux system is a powerful and high-performance multi-platform compiler launched by GNU. It is one of GNU's representative works. GCC is a super compiler that

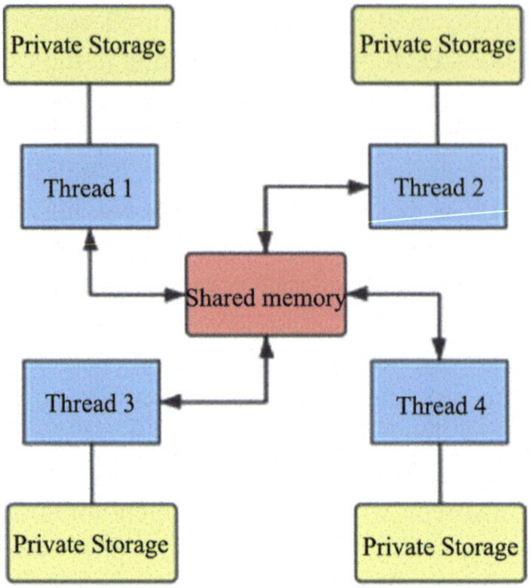

Fig. 3.10 Shared storage features

collection) compiler under the Linux platform and the Intel Visual Fortran compiler (embedded into the Visual Studio development environment) under the Windows platform.

OpenMP adopts a relaxed-consistency shared storage model, as shown in Fig. 3.10. Threads can use private storage space to "cache" their data without having to maintain precise real-time consistency of shared memory data at all times. Shared storage is a space where all OpenMP threads can access data. Each thread has its own temporary view of storage. This temporary view does not need to be part of the OpenMP storage model, but can be mapped to registers, caches, or local storage in the thread and shared storage space. Thread private storage is a storage space that can only be accessed by the local thread and not by other threads.

In the three-dimensional high-fidelity numerical method developed in this article, there are modules in the loop body that calculate the iterative solution of repetitive DO loops in the three directions of the coordinate system I, J, and K. These loop modules require a large amount of calculation, but the data are not related to each other. There is no direct interdependence, making it suitable for OpenMP thread-level parallelism. During program implementation, you only need to add corresponding compilation instruction statements (such as !$omp parallel do private (j)) to achieve multi-thread parallelism. During program execution, you can set system environment variables or call related library functions. Specify the number of threads to execute in parallel.

Figure 3.11 gives a schematic diagram of the OpenMP execution model. In the initial stage, a single thread (initial thread) is used for serial execution. For example, when encountering the three-direction loop iteration module of I, J, and K, multi-threading is used OpenMP calculates loop bodies in three directions simultaneously.

---

can compile executable programs on a variety of hardware platforms. Its execution efficiency is 20–30% higher than that of ordinary compilers on average.

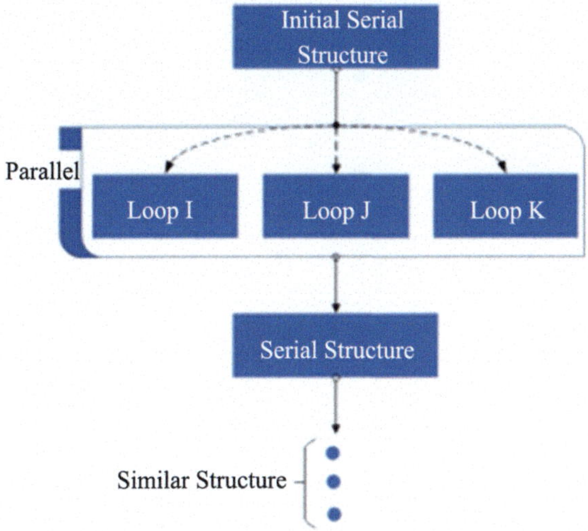

Fig. 3.11 OpenMP execution model

When all threads in the thread group reach the end of the parallel region, only the initial thread continues to execute, and other threads sleep or cancel. The parallel method based on OpenMP can improve the calculation speed of the original serial program to a certain extent, but since serial still accounts for the majority of the program, the ability to improve is limited.

3.6.3 MPI-Based Parallelism

For large-scale computing problems, due to the huge number of grids, the use of serial computing is not only unacceptable in engineering terms, but also places very high requirements on the storage space of a single process. Compared with thread-based OpenMP, process-based MPI has higher parallel efficiency. MPI is based on message passing to write parallel programs.

When implementing MPI parallel computing, it is considered that the solution process performed on each grid block is exactly the same, and the only difference is the boundary conditions. Therefore, the calculation task can be split, and each process is only responsible for the calculation of a single or a group of grid blocks. After each iteration is completed, MPI will Physical quantity information is transferred to each other. For a two-dimensional grid consisting of four grid blocks A, B, C, and D as shown in Fig. 3.12a (the green solid line in the figure represents the physical boundary, the red solid line Represents the inner boundary), when performing MPI parallel computing, these four grid blocks are calculated independently on four processes. The physical boundaries can be calculated directly within the respective processes according to the content of Sect. 3.5 ; but for the internal boundaries between blocks, message passing as shown in Fig. 3.12b is required. During data transfer, a virtual

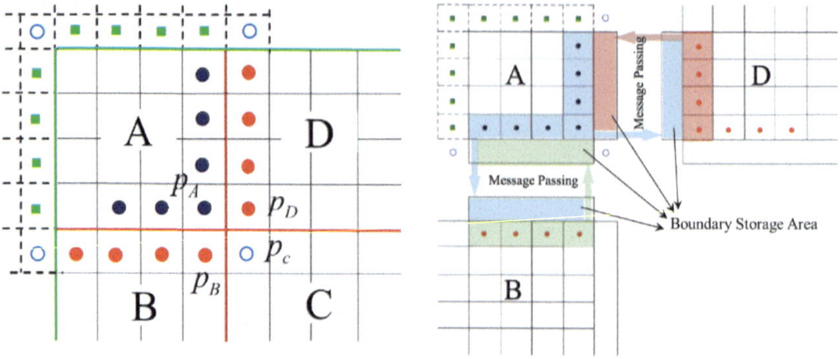

(a) Schematic diagram of the correlation between different boundaries and grid blocks

(b) Schematic diagram of message passing

Fig. 3.12 Schematic diagram of message passing (taking block A as the research object). Green square: physical boundary discrete point; red/black solid point: inner boundary discrete point; blue hollow point: corner point

grid (boundary storage area in the figure) is opened around grid block A, and the boundary discrete point information (red solid points) between adjacent grid blocks B and D adjacent to A is transferred through MPI communication. To the virtual grid, similarly, the information of the boundary discrete points (black solid points) within grid block A also needs to be transferred to the virtual grids corresponding to grid blocks B and D through MPI communication. Since the diagonal grid block C has no direct adjacent relationship with the grid block A, the value of the corner point (blue open circle) P_c on the grid block C cannot be obtained directly and needs to Interpolation approximation with the help of surrounding points. The first-order interpolation of points P_c in Fig. 3.12a is:

$$p_C = p_D + p_B - p_A \tag{3.25}$$

To minimize the consumption of memory and CPU time by MPI information transfer, the number of virtual grids should be as small as possible. For the second-order central difference format, at least one virtual grid is required. For the upwind format, the number of virtual grids required mainly depends on the reconstruction format. Table 3.2 gives different reconstruction formats and required virtual grids. The number of grids.

Many iterative calculations are performed independently on each process. When boundary (inner boundary and periodic boundary) physical field information is required, information is transferred between processes; when global flow field parameter calculations are performed, such as global residuals, progress Export flow, etc., the values of all processes are reduced and broadcast to all processes through collective communication; the control parameters of the main process and the solved full

Table 3.2 Minimum number of virtual meshes required for different reconstruction formats

Refactor format	Number of virtual grids
Zero-order reconstruction	1
Van Albada Limiter	1
Min mod limiter	1
Level 3 MUSCL	2
Level 5 WENO	3

flow field parameters are transmitted to all other processes through broadcasting. The main frame of the parallel program is shown as Fig. 3.13.

The calculation flow of the program is:

(1) Perform MPI initialization, specify the number of processes required for calculation, and number each process. The process with My_id = 0 is the "main

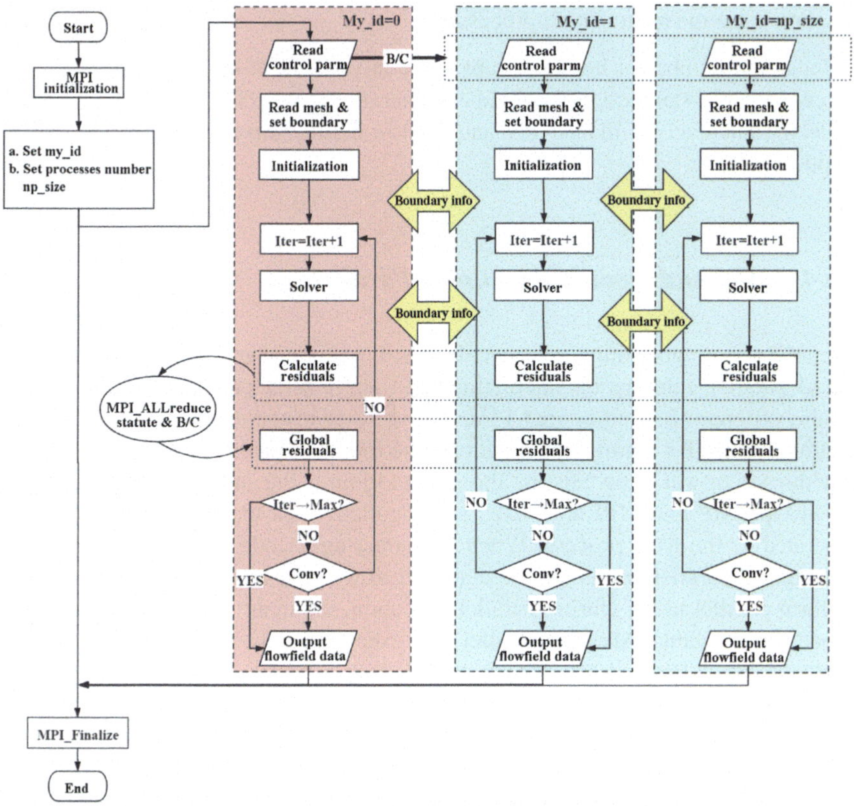

Fig. 3.13 Program flow chart—main frame

process" for reading control parameters and calculating full flow field parameters; The main process reads the control parameters and passes the control parameters to each process with My_id = 1 ~ np_size through broadcast;

(2) Each process reads the grid and boundary settings that it needs to process, and then initializes them, including the grid center point, the surface vector and volume of the grid unit, the distance to the nearest wall, the given initial value of the physical field, and variable initialization. Then the physical information at the inner boundary and periodic boundary is transferred to the connection location through MPI message sending and reception;

(3) The number of iteration steps "Iter" is counted to solve the flow field, including solving the mass equation, momentum equation, energy equation and turbulence model. Iterative updates of physical information on internal boundaries and periodic boundary conditions are still required, and MPI communication is still needed for information transfer;

(4) Calculate the residuals on each process, perform reduction operations through communication sets to obtain the maximum value of the entire flow field, and broadcast the results to all processes.

Judge the number of iteration steps and convergence. If the calculation reaches the convergence standard or the number of iteration steps reaches the set maximum value, the calculation will be exited and the flow field results on each process will be output.

3.6.4 MPI and OpenMP Coupled Parallelism

OpenMP is a programming interface designed to guide shared memory and multi-thread parallel compilation instructions and application programs when writing parallel programs for multi-core CPUs. It can be used in conjunction with C/C++ and Fortran. OpenMP's parallel model adopts the fork-join method, where fork creates new threads or wakes up existing threads, and join is the sum of multiple threads. When OpenMP and MPI are mixed and parallelized, a main thread will first be generated in the MPI process. When encountering a code segment that requires fine-grained parallel execution, the main thread will spawn multiple sub-threads to perform parallel tasks. During parallel execution, the main thread and the derived thread work together. After the parallel code execution ends, the derived thread exits or blocks and no longer participates in calculations, while the main thread takes over the control flow and continues execution.

Figure 3.14 programming structure using MPI and OpenMP coupling is given. First, the grid of the computational domain is reasonably allocated to multiple processes to achieve coarse-grained parallelism. During the solving process on each process, when encountering When do solves algebraic equations in multiple dimensions, the main thread in MPI forks out multiple threads for fine-grained parallelism. After the algebraic equation is solved, all parallel threads join back to the main thread,

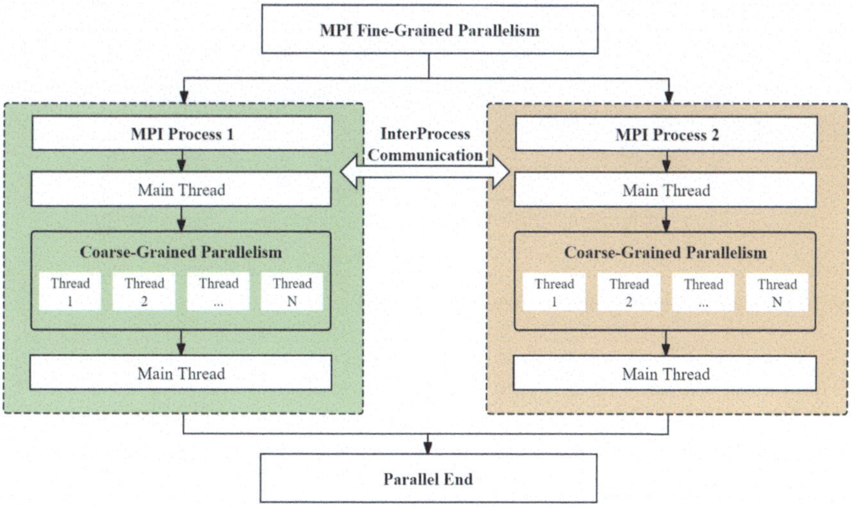

Fig. 3.14 MPI and OpenMP coupled programming

and the main thread performs subsequent work. For the computing tasks assigned to each MPI process, the OpenMP multi-threading mechanism can be used to speed up the solution of each discrete algebraic equation system.

3.6.5 Parallel Method Verification and Testing

1. **OpenMP Test**

Based on the RANS method, this chapter develops the thread-level OpenMP parallel method. This section evaluates the speed of program operation before and after parallelization, and finds the optimal number of threads in order to achieve efficient utilization of hardware. The three systems tested are:

(1) Linux Virtual Machine: The computer processor is Intel(R) Core(TM) i7-4790K, and the CPU frequency is 4.0 GHz;
(2) Alibaba Cloud ECS Linux System: Alibaba Cloud ECS server 4-core 8 GB Series III shared computing type;
(3) Windows System: The computer processor is Intel(R) Core(TM) i7-4790K, and the CPU frequency is 4.0 GHz.

Record the time it takes to iteratively calculate 100 steps with a data volume of 50,000 grids. The results are shown in Fig. 3.15. Triangular points, square points, and dots represent Linux virtual machines, Alibaba Cloud ECS Linux systems, and Windows respectively. System operation results. From the system level comparison, the Windows system used in this article's test runs the fastest. From the perspective of

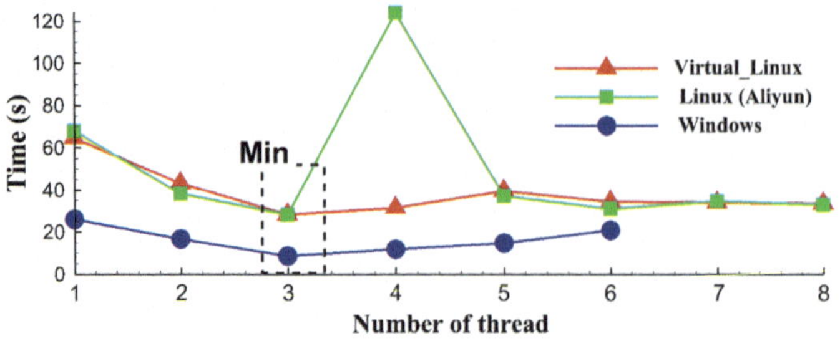

Fig. 3.15 Optimal number of threads based on OpenMP

the number of threads (Fig. 3.15, the abscissa represents the number of threads used), using 3 threads takes the shortest time to complete the same number of calculations. Therefore, using OpenMP parallel computing can be achieved given the number of 3 threads. Optimal utilization of hardware.

2. **MPI Verification and Testing**

(1) Accuracy verification of data transfer between MPI parallel processes

In order to check the accuracy of data transfer between multiple processes in parallel based on MPI, this article compares parallel and serial results based on the RANS method. Taking the controllable diffusion cascade NPU-A1 as the research object, AutoGrid™ is used for grid division, and the O4H type grid topology is used to divide the calculation domain into 5 grid block structures, such as Fig. 3.16 is shown. Only one grid operation is performed on each process, and each process has message communication with at least three processes. The blade surface and lower end wall are given adiabatic no-slip boundaries, the upper end wall is given slip boundary condition, and both sides of the flow channel are given periodic boundary conditions.

Serial programs and MPI-based parallel programs are used for numerical simulation. Since this section focuses on verifying the consistency of the calculation results of parallel and serial programs, this section does not discuss the accuracy of numerical results compared with experiments. The convergence curves of the inlet and outlet mass flow rates of a single channel during the iterative process are shown in Fig. 3.17. It can be seen that the curve convergence process during the iterative process is exactly the same, and the inlet and outlet flow rates can be guaranteed to be consistent after convergence. The flow field at 25% of the blade height is selected for comparison. The dimensionless static pressure results are shown in Fig. 3.18. The left and right sides are the serial and MPI parallel results respectively. It can be seen from the static pressure contour distribution, the size and distribution of static pressure values agree well, verifying the accuracy of message transmission between MPI processes.

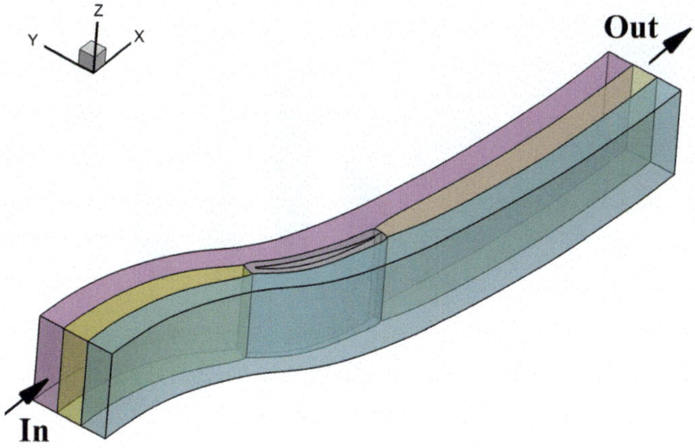

Fig. 3.16 Computational grid topology

Fig. 3.17 Comparison of serial and parallel iterative processes—inlet and outlet mass flow rate

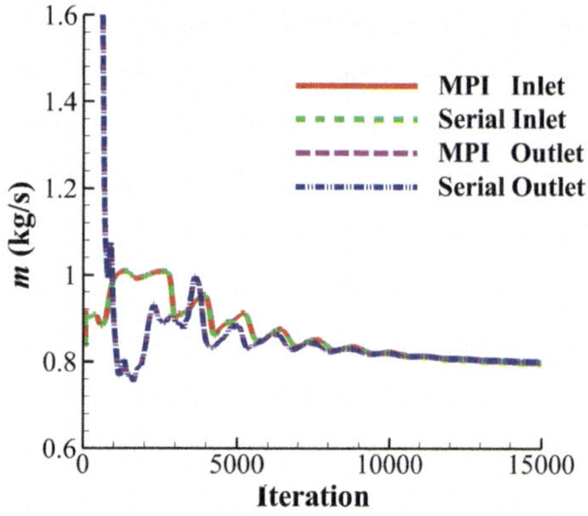

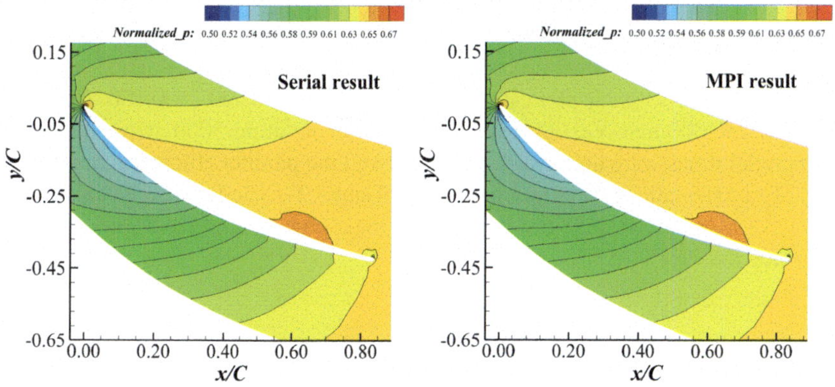

Fig. 3.18 Comparison of serial and parallel calculation results—dimensionless static pressure comparison

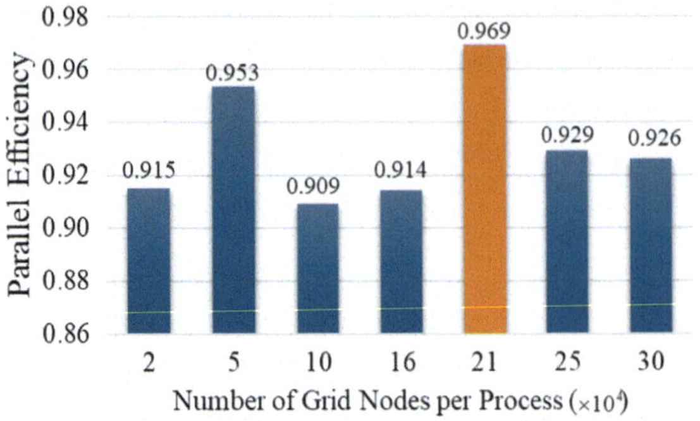

Fig. 3.19 Program parallel efficiency test results

(2) MPI parallel efficiency test

Parallel efficiency is an evaluation index for measuring the performance of parallel programs. Before performing parallel computing, it is necessary to test the parallel efficiency in order to achieve the most efficient use of hardware. Parallel efficiency $Eff(n)$ is defined as:

$$Eff(n) = \frac{S(n)}{n} \tag{3.26}$$

In the formula:

n Number of parallel computing processes

$S(n)$ Speedup ratio.

The speedup is the ratio of serial time t_{serial} to parallel time $t_{parallel}$, defined as:

$$S(n) = \frac{t_{serial}}{t_{parallel}} \tag{3.27}$$

The MPI calculation in this section uses 3 processes for testing. The same grid amount is given to each process, which are 20,000, 50,000, 100,000, 160,000, 210,000, 250,000 and 300,000 respectively. After iterating 5000 steps, the serial and parallel times were counted respectively, and the parallel efficiency was calculated. Test calculations were completed on "Tianhe-2". The test results are shown in Fig. 3.19. About 200,000 grid nodes on a single process have the highest parallel efficiency.

References

1. Li Xuesong, Gu Chunwei. Simulation study of separation vortex of high-pressure stage stator blades of compressors with large separation. Journal of Engineering Thermophysics, 2009, 30(01): 31–34.

2. Chen Meilan, Li Xuesong, Gu Chunwei. Research on DES and RANS models in compressor cascade flow. Shenzhen: Large Aircraft Key Technology High-Level Forum and China Aeronautical Society 2007 Academic Annual Conference, 2007.
3. Yang Zeyu. Application of DDES method in compressor plane cascade. Xi'an: Northwestern Polytechnical University, 2016: 53–68.
4. Liu Y, Yan H, Lu L. Numerical study of the effect of secondary vortex on three-dimensional corner separation in a compressor cascade. International Journal of Turbo & Jet-Engines, 2016, 33(1): 9–18.
5. Liu Ruoyang, Hou Anping, Shan Shujun, etc. Numerical study of unsteady flow of cascade separation vortex based on DDES method. Propulsion Technology, 2017(01): 16–26.
6. Gu Chunwei, Feng Fan, Li Xuesong, et al. Discussion on the application of DES model in compressor subsonic rotor. Journal of Engineering Thermophysics, 2008, 29(6): 951–956.
7. Li Xuesong, Feng Fan, Gu Chunwei. Study on separated vortex simulation of static and dynamic interference flows. Journal of Engineering Thermophysics, 2009(06): 953–956.
8. Hongsik I M, Chen X, Zha G. Detached-eddy simulation of rotating stall inception for a full-annulus transonic rotor. Journal of Propulsion and Power, 2012, 28(4): 782–798.
9. Hongsik I M, Chen X, Zha G. Detached eddy simulation of transonic rotor stall flutter using a fully coupled fluid-structure interaction. Vancouver: Proceedings of the ASME 2011 Turbo Expo, 2011.
10. Gan J Y, Im H S, Zha G C. Simulation of stall inception of a high speed axial compressor with rotor-stator interaction. Orlando: 51st AIAA/SAE/ASEE Joint Propulsion Conference, 2015.
11. Gan J Y., Im H S, Zha G C. Delayed detached eddy simulation of rotating stall for a full annulus transonic axial compressor stage. Seoul: Proceedings of ASME Turbo Expo 2016, 2016.
12. Shi K, Fu S. Study of shock/blade tip leakage vortex/boundary layer interaction in a transonic rotor with IDDES method. San Antonio: Proceedings of ASME Turbo Expo 2013, 2013.
13. Yamada K, Kikuta H, Iwakiri K, et al. An explanation for flow features of spike-type stall inception in an axial compressor rotor. Journal of turbomachinery, 2013, 135(2): 1–11.
14. Kusano K, Jeong J H, Yamada K, et al. Detached eddy simulation of unsteady flow field and prediction of aerodynamic sound in a half-ducted propeller fan. Hamamatsu: Proceedings of the ASME-JSME-KSME 2011 Joint Fluids Engineering Conference, 2011.
15. Yamada K, Furukawa M, Nakakido S, et al. Large-scale DES analysis of unsteady flow field in a multi-stage axial flow compressor at off-design condition using k computer. Montreal: Proceedings of the ASME Turbo Expo 2015, 2015.
16. Ma W. Experimental investigation of corner stall in a linear compressor cascade. Lyons: University of Lyon, 2012: 106–187.
17. Ma W, Gao F, Ottavy X, et al. Numerical investigation of intermittent corner separation in a linear compressor cascade. Seoul: Proceedings of ASME Turbo Expo 2016, 2016.
18. Liu Y, Yan H, Lu L, et al. Investigation of vortical structures and turbulence characteristics in corner separation in a linear compressor cascade using DDES. Journal of Fluids Engineering, 2017, 139(2): 1–14.
19. Li R, Zhao L, Ge N, Gao L, Ni M. Grid resolution assessment method for hybrid RANS-LES in turbomachinery. Engineering Applications of Computational Fluid Mechanics, 2021, 16(1): 279–295.
20. Gao, L., Li, R., Miao, F., Cai, Y. Unsteady Investigation on Tip Flow Field and Rotating Stall in Counter-Rotating Axial Compressor. Journal of Engineering for Gas Turbines and Power, 2015, 137(7): 1–11.
21. Lau, Y. L., Leung, R. C. K., So, R. M. C. Vortex-induced vibration effect on fatigue life estimate of turbine blades. Journal of Sound and Vibration, 2007, 307(3–5): 1–22.
22. Sharma, S., Broatch, A., García-Tíscar, J., Allport, J. M., Nickson, A. K. Acoustic characterization of a small high-speed centrifugal compressor with casing treatment: An experimental study. Aerospace Science and Technology, 2019, 95: 1–15.
23. Li R, Gao L, Zhao L, Ma C, Lin S. Dominating unsteadiness flow structures in corner separation under high Mach number. AIAA Journal, 2019, 57(7): 2923–2932.

24. Li, Y. H., Wu, Y., Zhou, M., Su, C. B., Zhang, X. W., Zhu, J. Q. Control of the corner separation in a compressor cascade by steady and unsteady plasma aerodynamic actuation. Experiments in Fluids, 2010, 48(6): 1015–1023.
25. Gourdain, N. (2015). Prediction of the unsteady turbulent flow in an axial compressor stage. Part 2: Analysis of unsteady RANS and LES data. Computers and Fluids, 106: 119–129.
26. Spalart, P. R., Jou, W. H., Strelets, M. K., Allmaras, S. R. Comments on the feasibility of LES for wings and on a hybrid RANS/LES approach. Baton Rouge: Proceedings of the First AFOSR International Conference on DNS/LES, 1997.
27. Breuer, M., Jovičić, N., Mazaev, K. Comparison of DES, RANS and LES for the separated flow around a flat plate at high incidence. International Journal for Numerical Methods in Fluids, 2003, 41(4): 357–388.
28. Riéra, W., Marty, J., Castillon, L., Deck, S. Zonal detached-eddy simulation applied to the tip-clearance flow in an axial compressor. Journal of Aircraft, 2016, 53(4): 2377–2392.
29. Spalart P R. Young-person's guide to detached-eddy simulation grids. NASA Langley Technical Report: NASA, 2001: 1–23.
30. Roe P L. A survey of upwind differencing techniques. Williamsburg: 11th International Conference on Numerical Methods in Fluid Dynamics, 1988.
31. Zhou Yu, Yan Chao. Performance analysis of different types of entropy correction in Roe format. Journal of Beihang University, 2009, 35(3): 356–360.
32. NASA Langley Research Center. Turbulence modeling resource_the spalart-allmaras turbulence model. NASA, 2019, https://turbmodels.larc.nasa.gov/spalart.html.
33. Oden JT, Belytschko T, Fish J, et al. Simulation-based engineering science: Revolutionizing engineering science through simulation. Report of the National Science Foundation Blue Ribbon Panel on Simulation-Based Engineering Science: NSF Engineering Advisory Committee, 2006: 1–34.
34. Keyes D, Colella P, Dunning Jr T, et al. A science-based case for large-scale simulation. Technical Report of US Department of Energy: Office of Science U.S. Department of Energy, 2003: 1–76.
35. Gorrell SE, Tsung FL, Yao J, et al. Computational science and engineering advances understanding of complex unsteady flows in high performance fans and compressors. Denver: Proceedings of the HPCMP Users Group Conference 2006, 2006.
36. Mavriplis DJ, Darmofal D, Keyes D, et al. Petaflops opportunities for the NASA fundamental aeronautics program. Miami: 18th AIAA Computational Fluid Dynamics Conference, 2007.
37. Flynn MJ. Very high-speed computing systems. Proceedings of the IEEE, 1966, 54(12): 1901–1909.
38. Liao XK, Xiao LQ, Yang CQ, et al. Milkyway-2 supercomputer: system and application. Frontiers of Computer Science, 2014, 8(3): 345–356.
39. Yang XJ, Liao XK, Lu K, et al. The TianHe-1A supercomputer: its hardware and software. Journal of Computer Science and Technology, 2011, 26(3): 344–351.
40. Clarke L, Glendinning L, Hempel R. The MPI Message Passing Interface Standard: Monte Verità: Programming Environments for Massively Parallel Distributed Systems, 1994.

Chapter 4
High-Fidelity Flow Field Massive Data Analysis Method for Compressors

4.1 Introduction

The high-fidelity flow fields generated by DES-series methods exhibit a distinctive characteristic of high spatiotemporal resolution. In the spatial domain, extensive data from grid nodes enable the capture of refined flow field details. In the temporal domain, finely-resolved unsteady physical fields enable the capture of a broad frequency spectrum. Given the multi-scale, broadband spectrum, and highly nonlinear coupling inherent in compressor flow fields, tackling the complexities of massive unsteady flow field data presents formidable challenges. These include extracting primary frequency features, identifying dominant vortex structures, discerning causal relationships within a plethora of turbulent structures, and deciphering the intertwined evolution of multiple vortex structures. This chapter introduces methods for analyzing massive unsteady flow field data, including frequency domain analysis, vortex identification techniques, as well as Proper Orthogonal Decomposition (POD) and Dynamic Mode Decomposition (DMD). The synergy and integration of these methods promise to enhance the efficacy of utilizing massive data in research endeavors.

4.2 Frequency Domain Analysis for Unsteady Flow Fields

4.2.1 Frequency Domain Analysis

The spectral analysis of the transient flow of a compressor can extract the crucial flow parameters such as amplitude, frequency, and phase, which facilitates the identification of dominant frequencies at specific locations, aids in the determination of critical influencing factors for performance and enables more effective utilization of unsteady flow phenomena. The time-domain description of signal means the collected

R. Li et al., *Detached-Eddy Simulation Series Methods for Compressor Internal Flow*,
https://doi.org/10.1007/978-981-96-0255-1_4

signal with time as independent variable. However, elucidating the frequency-domain structure and the magnitudes of individual frequency components can sometimes prove challenging. Therefore, in dynamic testing scenarios, it becomes necessary to use frequency-domain description methods, where frequency serves as the independent variable, revealing the relationships of signal components among magnitudes, phases, and frequency. The three variables can be transformed into one another through mathematical operations, all representing the same measured signal. The spectral representation of dynamic data can be obtained via the fast Fourier transform, which converts a time-domain function into a frequency-domain function and decomposes a complex waveform into a sum of sinusoids of various frequencies. This mathematical relationship can be formally expressed as [1]:

$$G(f) = \int_{-\infty}^{0} g(t)e^{-j2\pi ft}\mathrm{d}t \tag{4.1}$$

where, $g(t)$ represents the dynamic signal, which is decomposed into a sum of sinusoidal functions, while $G(f)$ represents the Fourier transform of $g(t)$ and $j = \sqrt{-1}$. The discrete form of the Fourier transform, commonly referred to DFT, is expressed as follows:

$$Y_k = \sum_{n=0}^{N-1} g_n e^{-\frac{j2\pi kn}{N}} \quad (k = 0, 1, 2, \ldots, N-1), \tag{4.2}$$

where Y_k represents the Fourier amplitude and $g_n = g(n\tau)$.

The FFT algorithm is a rapid computational technique employed to compute the DFT. It partitions the complete data sequence $\{g_n\}$ into shorter segments for DFT computation, substituting the need to directly compute the DFT of the original sequence. By calculating the DFT of these shorter segments and subsequently merging them, the DFT of the entire sequence $\{g_n\}$ can be obtained efficiently. Utilizing the FFT transformation facilitates the generation of the frequency spectrum of dynamic pressure.

There exists an example using fluctuation data of outlet flow varying with time from a compressor rotor to demonstrate the application of spectral analysis method in understanding compressor flow fields. Figure 4.1 depicts the convergence history of outlet flow under different operating conditions during numerical simulation. Each iteration step in the numerical simulation corresponds to a physical time step, hence Fig. 4.1 can also be considered a time-domain representation of outlet flow fluctuations.

Figure 4.1a–d illustrate four typical operating conditions ranging from the design condition to approaching surge as the flow rate decreases gradually. Figure 4.2 presents the frequency-domain results of outlet flow fluctuations under these four operating conditions. At the design condition, the frequency spectrum of flow fluctuations reveals only the presence of the 2BPF, which represents the frequency of

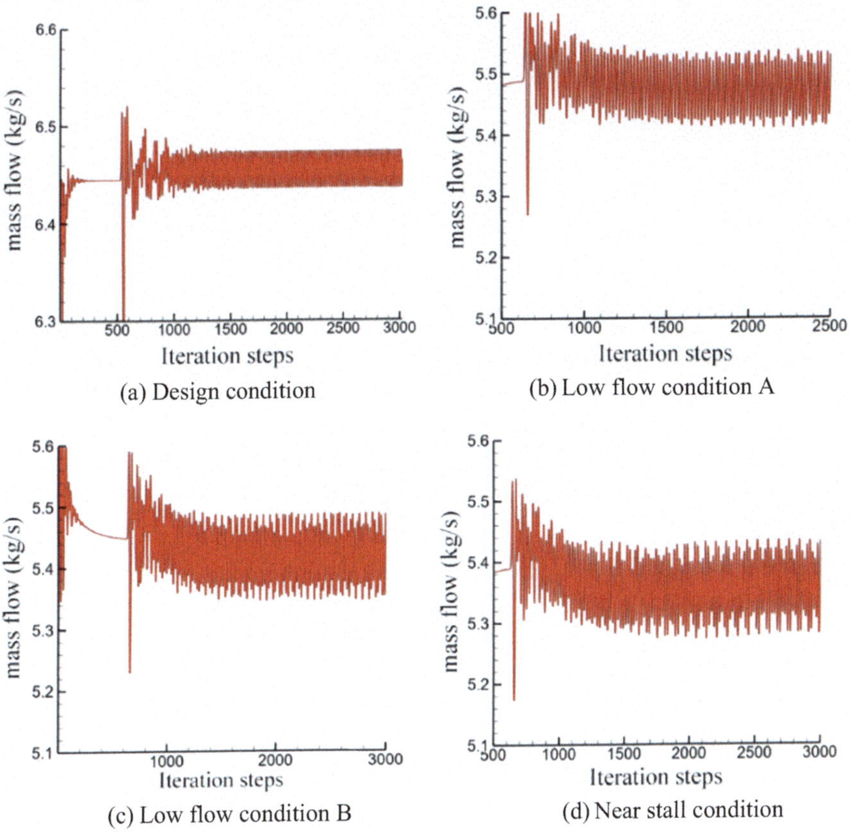

Fig. 4.1 The chart of outlet flow convergence under different conditions

mutual interference between rotor row. As the flow rate decreases, the frequency spectrum of flow fluctuations at low-flow conditions A and B reveals the emergence of the 0.64BPF as Fig. 4.2b, c shows. Upon approaching surge, this characteristic frequency abruptly increases to become 1.5BPF as Fig. 4.2d shows. Since this change occurs around the surge condition, it can be utilized as a surge warning signal. The variation of characteristic frequencies in the flow field implies changes in the flow field structure. Thus, spectral analysis can aid in identifying operating conditions where changes in flow field structures occur. Further analysis involves placing numerical monitoring points at different locations in the flow field, where the magnitude of fluctuations can assist in identifying the source of fluctuations corresponding to characteristic frequencies in the flow field.

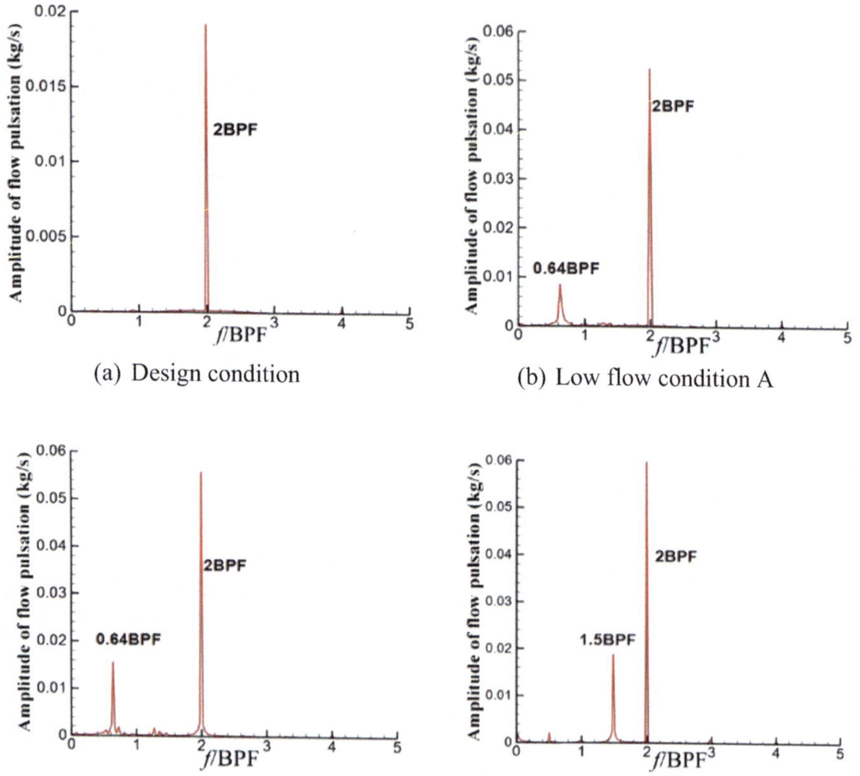

(a) Design condition (b) Low flow condition A

(c) Low flow condition B (d) Near stall condition

Fig. 4.2 Spectrum chart of outlet flow under different working conditions (BPF is blade passing frequency) [2]

4.2.2 *Stall Mass Characterization*

Within the operational scope of a compressor, the rotor passing frequency (RPF) and blade passing frequency (BPF) stand as pivotal frequency characteristics of the flow regime.

$$\text{RPF} = \frac{n}{60} \tag{4.3}$$

$$\text{BPF} = N \cdot \frac{n}{60}, \tag{4.4}$$

where n represents the rotational velocity and N denotes the count of rotor blades.

As the compressor nears the threshold of unstable operation, the occurrence of rotating stall cells becomes prominent, manifesting as low-pressure regions propagating circumferentially from the leading edge of the blade tip. The characterization of stall cells is primarily described through parameters such as the number of stall cells N_0 and their circumferential propagation velocity w_1.

$$N_0 = \frac{T_{cr}}{T_{osc}}$$

$$\omega_1 = N_1 \cdot \omega_2 / N_2 \tag{4.5}$$

where, T_{cr} represents the rotational period of the stall cell acquired from pressure signals measured by all sensors circumferentially, T_{osc} represents the rotational period of an individual stall cell obtained from a single sensor, ω_2 represents the rotor rotational speed, and N_1 signifies the number of rotor rotations completed upon the completion of N_2 stall cell rotations.

An illustrative demonstration of stall cell characteristic analysis method is presented, employing unsteady pressure data obtained from the inner casing surface adjacent to the leading edge of compressor blades. The dataset is from experimental trials conducted on a low-speed axial compressor. Seven dynamic pressure sensors are strategically positioned at locations P2-1 to P2-7 in Fig. 4.3. These sensors, which are positioned proximally to the blade leading edge, synchronously capture unsteady pressure fluctuations across various operational conditions. Specifically, the analysis focuses on two high-flow conditions and two low-flow conditions, designated as depicted in Fig. 4.4. Measurements are performed for conditions labeled 2, 9, 13, and 14.

In Fig. 4.5, the dynamic pressure signals from sensors P2-1 to P2-7 at simultaneous time instances are depicted. Under high-flow conditions, as denoted by points CON-2 and CON-9, the signals exhibit a relatively smooth profile without

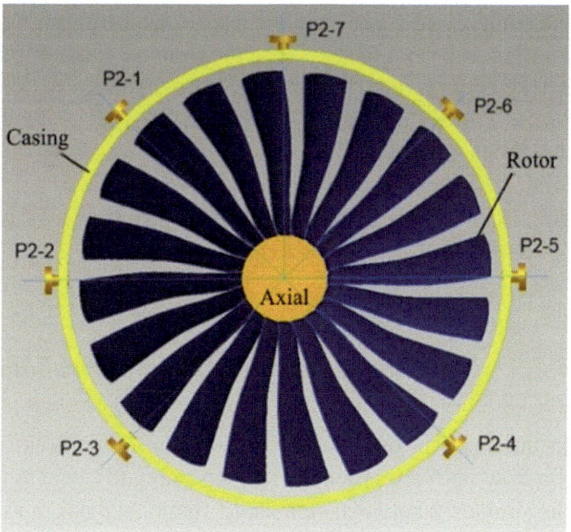

Fig. 4.3 Static pressure measurement locations near the leading edge of the compressor blades (P2-1 to P2-7)

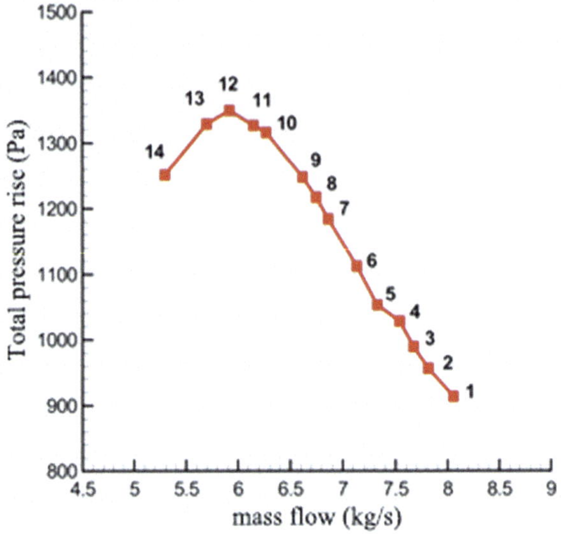

Fig. 4.4 The compressor pressure rise characteristic curve at 2500 rpm

discernible perturbations. However, as the compressor approaches the vicinity of surge condition, indicated by point CON-13 where the outlet flow area decreases and the load increases, the amplitude of pressure data across all sensors markedly rises. Additionally, irregular "spikes" indicates the heightened unsteady pulsations in the tip clearance flow, which suggests the compressor's gradual approach towards an unstable operational regime. At the surge point, denoted as CON-14, stable stall cells have formed circumferentially. This is evident in the dynamic pressure signals as periodic fluctuations in pressure characteristics propagating from P2-1 to P2-7.

Due to the inherent decrease in pressure accompanying surge, regions within the compressor surge pressure field exhibiting significantly low pressure are termed rotating stall cells [3–5]. Upon the onset of surge conditions, precisely one stall cell is formed within the compressor. As evidenced by the completion of approximately $N_1 = 7/8$ cycle from P2-1 to P2-7, the stall cell rotation is observed. Analyzing the irregular "spikes" corresponding to the abscissa of P2-1 to P2-7 in the graph indicates the rotor has completed roughly 1.88 of a cycle. Therefore, in accordance with Eq. (4.5), the circumferential propagation velocity of the stall cell is computed to be 46.5% of the rotor speed (Fig. 4.5).

4.2.3 Correlation Coefficients and Probability Statistics

In the analysis of unsteady flow fields in compressors, the periodic variations of the flow field are crucial reference standards, which are typically characterized by the similarity between signals. In signal analysis, correlation analysis is commonly employed to assess the similarity of signals, reflecting the linear relationship between

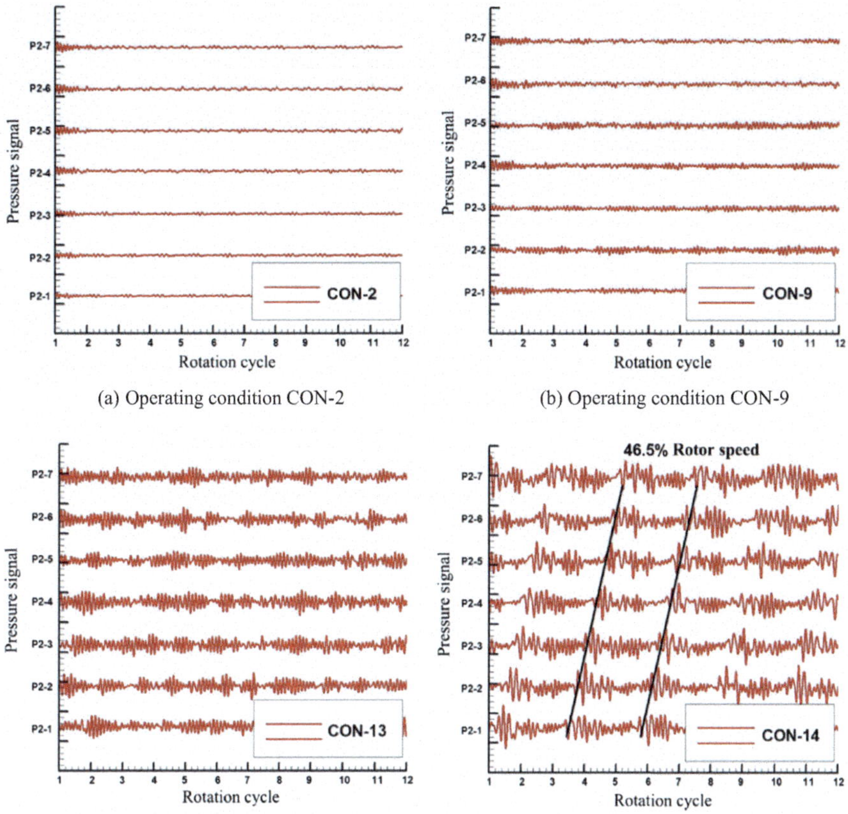

(a) Operating condition CON-2

(b) Operating condition CON-9

(c) Operating condition CON-13

(d) Operating condition CON-14

Fig. 4.5 Time-domain signals of various sensors after filtering under different operating conditions

the same or different signals at different time instances. Correlation analysis algorithms are much simpler than FFT and are more straightforward for online fast computations. Therefore, this method is frequently utilized in surge detection during aircraft engine testing. The calculation formula is depicted in Eq. (4.6).

$$R(t) = \frac{\sum_{i=t-wnd}^{i} (\overline{p}(n,i) \cdot \overline{p}(n, i - shaft))}{\sqrt{\sum_{i=t-wnd}^{i} \overline{p}(n,i)^2 \cdot \sum_{i=t-wnd}^{i} \overline{p}(n, i - shaft)^2}} \tag{4.6}$$

The correlation calculation principle entails the segmentation of pressure signals $\overline{p}(n, i), \overline{p}(n, i - shaft)$ at corresponding time points from the preceding and subsequent two cycles, where *shaft* is defined as the specific number of points within one revolution. Within a designated calculation window (*wnd*), the correlation coefficient R is computed for each data point. R, ranging from -1 to 1, serves as a measure of signal correlation, with values closer to 1 indicating stronger periodicity and thus better correlation. As the compressor nears its unstable boundary, the periodicity of the tip clearance flow is disrupted, resulting in a gradual reduction in R.

Autocorrelation analysis is commonly utilized to assess the correlation of signals at the same spatial position but different temporal instances, which serves to quantify the degree of correlation of the tip clearance flow under varying operating conditions. Conversely, for evaluating the correlation of signals at distinct spatial positions, cross-correlation analysis is typically employed, which provides insights into the degree of correlation of the tip clearance flow at different locations under identical operating conditions.

In the analysis of the correlation coefficient R, it has been observed that although R exhibits a trend corresponding to changes in flow rate, the stochastic nature of signal acquisition and contamination by noise lead to random fluctuations in R. Consequently, determining the precise magnitude of R becomes challenging. To address this issue, the concept of probability statistical distribution functions is introduced for the analysis of the correlation coefficient. In this context, F represents the probability distribution value, while x denotes the probability calculation threshold.

$$F(R) = f(R < x) \tag{4.7}$$

The interpretation of the probability distribution function is: the abscissa denotes the values of the correlation coefficient R, while the ordinate signifies the probability value of the data points containing values less than a certain correlation coefficient. A curve that predominantly shifts towards the right side of the graph, indicating higher values along the abscissa, suggests that the correlation coefficient tends closer to 1. This trend signifies a more intact periodicity of the flow field.

Taking the experimental measurements of the compressor rotor from Fig. 4.3 as a reference, Fig. 4.6 illustrates the probability distribution of autocorrelation coefficients at 7 circumferential positions for four distinctive operating conditions. The 7 circumferential measurement positions are indicated by P2-1 to P2-7 in Fig. 4.3. The four characteristic operating conditions correspond to those depicted in Fig. 4.4, specifically, conditions prior to entering surge (CON-2 and CON-9) and conditions near surge (CON-13 and CON-14).

Before the compressor enters surge conditions (CON-2 and CON-9), the pressure data measured at positions P2-3, P2-4, and P2-6 exhibit significantly higher probabilities of occurrence of small autocorrelation coefficients compared to the other positions. For instance, in the CON-2 condition (Fig. 4.6a), the probabilities of occurrence of autocorrelation coefficients below 0.85 at these three positions are 50%, 70%, and 70%, respectively, while at the other positions, this value is below 10%. When the compressor operates under surge conditions, such as CON-14 (Fig. 4.6d), the probability distribution curves at different positions overlap, indicating a consistent pattern in the distribution of autocorrelation coefficients. This suggests that although the autocorrelation coefficients at different positions exhibit different temporal variations, their probabilities of occurrence are equal, indicating consistency in the unsteady intensity of the flow field at various circumferential positions under surge conditions.

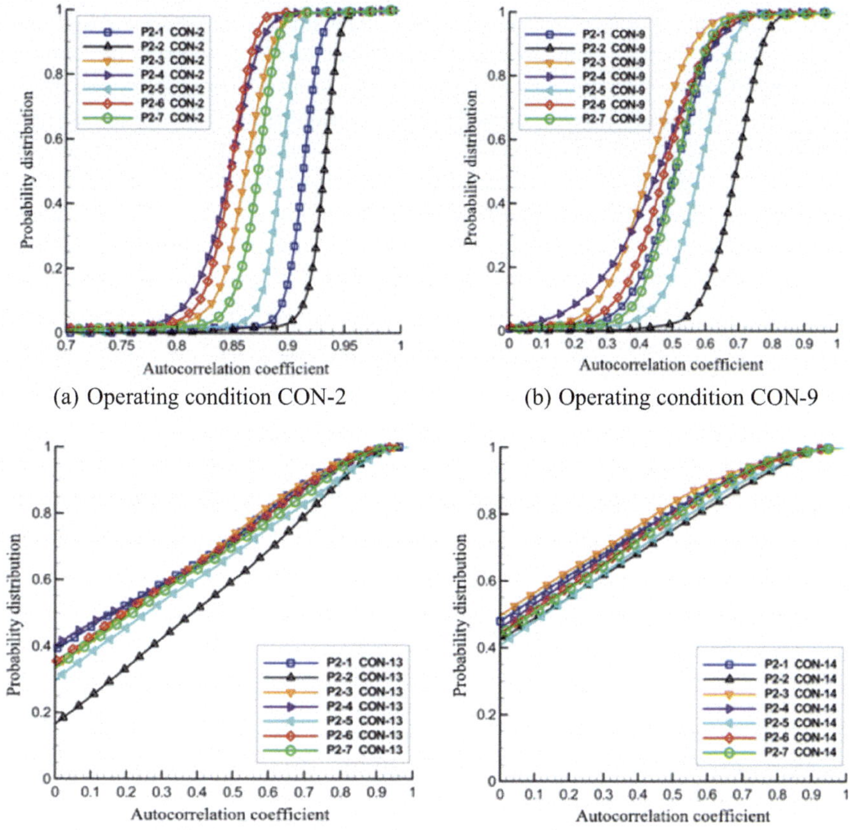

(a) Operating condition CON-2

(b) Operating condition CON-9

(c) Operating condition CON-13

(d) Operating condition CON-14

Fig. 4.6 Probability distribution of autocorrelation coefficients at characteristic operating points

4.3 Vortex Identification Methods in the Complex Flow Field

Vortex flows act as the "musculature" of a flow field; the spatial arrangement of vortex structures constitutes the framework of the flow. Therefore, capturing and identifying vortex structures within the flow field of a compressor is fundamental to the mechanisms of fluid flow. However, vortex identification itself is a challenging task. Despite this, the discernment of vortices presents a formidable challenge. Conceptually, vortices are indicative of the rotational motion within a fluid; it follows that the presence of fluid rotation is synonymous with the existence of vortices, and vice versa. However, to date, the academic community has not provided a strict mathematical definition of a vortex. In 1858, Helmholtz introduced the concepts of vortex filaments, vortex lines, and vortex tubes [6], using vorticity to represent vortex filaments. However, researches have shown that vorticity and vortices are two completely different concepts. For example, in a laminar shear layer, the vorticity is high, but it is not caused by rotational flow. In 1991, Robinson [7] found that the correlation between regions of high vorticity and actual fluid rotation is very low. DNS numerical simulations by Wang [8] and others also confirmed that vorticity is vastly different from the size and direction of vortices. In response to these complexities, the research community has since advanced vortex identification techniques predicated on eigenvalue-based criteria, exemplified by the Q criterion and the λ_2 criterion. These methods, as categorized by Liu [9], are considered second-generation vortex identification techniques. Furthermore, third-generation methods, such as those spearheaded by Liutex have also emerged [10]. This section will delve into an exposition of these principal methodologies and present case studies of their application within the context of compressor flow fields, showcasing their utility in the characterization and analysis of vortex structures.

The existence of vortices inevitably leads to variations in velocity. According to the Cauchy–Stokes decomposition theorem, the velocity gradient $\nabla \cdot \mathbf{W}$ can be decomposed into a symmetric part $\boldsymbol{A}$ and an antisymmetric part $\boldsymbol{B}$, which is expressed as:

$$\nabla \cdot \mathbf{W} = \mathbf{A} + \mathbf{B} = \frac{1}{2}\left(\nabla \cdot \mathbf{W} + (\nabla \cdot \mathbf{W})^{\mathrm{T}}\right) + \frac{1}{2}\left(\nabla \cdot \mathbf{W} - (\nabla \cdot \mathbf{W})^{\mathrm{T}}\right) \tag{4.8}$$

In a Cartesian coordinate system, the symmetric part $\boldsymbol{A}$ can be expressed as Eq. (4.9), which represents the fluid's stretching, compression, and symmetric deformation. The antisymmetric part B is represented by Eq. (4.10). According to the definition of vorticity $\psi = \nabla \times \mathbf{W}$, it represents the rigid body rotation.

$$\mathbf{A} = \begin{bmatrix} \frac{\partial w_x}{\partial x} & 0 & 0 \\ 0 & \frac{\partial w_y}{\partial y} & 0 \\ 0 & 0 & \frac{\partial w_z}{\partial z} \end{bmatrix} + \begin{bmatrix} 0 & \frac{1}{2}\left(\frac{\partial w_x}{\partial y} + \frac{\partial w_y}{\partial x}\right) & \frac{1}{2}\left(\frac{\partial w_x}{\partial z} + \frac{\partial w_z}{\partial x}\right) \\ \frac{1}{2}\left(\frac{\partial w_x}{\partial y} + \frac{\partial w_y}{\partial x}\right) & 0 & \frac{1}{2}\left(\frac{\partial w_y}{\partial z} + \frac{\partial w_z}{\partial y}\right) \\ \frac{1}{2}\left(\frac{\partial w_x}{\partial z} + \frac{\partial w_z}{\partial x}\right) & \frac{1}{2}\left(\frac{\partial w_y}{\partial z} + \frac{\partial w_z}{\partial y}\right) & 0 \end{bmatrix} \tag{4.9}$$

$$\mathbf{B} = \begin{bmatrix} 0 & \frac{1}{2}\left(\frac{\partial w_x}{\partial y} - \frac{\partial w_y}{\partial x}\right) & \frac{1}{2}\left(\frac{\partial w_x}{\partial z} - \frac{\partial w_z}{\partial x}\right) \\ \frac{1}{2}\left(\frac{\partial w_y}{\partial x} - \frac{\partial w_x}{\partial y}\right) & 0 & \frac{1}{2}\left(\frac{\partial w_y}{\partial z} - \frac{\partial w_z}{\partial y}\right) \\ \frac{1}{2}\left(\frac{\partial w_z}{\partial x} - \frac{\partial w_x}{\partial z}\right) & \frac{1}{2}\left(\frac{\partial w_z}{\partial y} - \frac{\partial w_y}{\partial z}\right) & 0 \end{bmatrix} \tag{4.10}$$

4.3.1 Q Criterion

The Q criterion, proposed by Hunt et al. [11], has been widely applied in the identification of vortex structures in flow fields [12, 13]. The Q criterion captures the essence of vortex structures by requiring not only the presence of vorticity, represented by the antisymmetric tensor B, but also that this vorticity must be sufficient to overcome the countervailing effects represented by the symmetric tensor A. The mathematical formulation is expressed as:

$$Q = \frac{1}{2}\left(\|\mathbf{B}\|_F^2 - \|\mathbf{A}\|_F^2\right) \tag{4.11}$$

In the equation, $\|\cdot\|_F^2$ denotes the Frobenius norm of the matrix. The expression in a Cartesian coordinate system is given by:

$$Q = -\frac{1}{2}\left(\left(\frac{\partial w_x}{\partial x}\right)^2 + \left(\frac{\partial w_y}{\partial y}\right)^2 + \left(\frac{\partial w_z}{\partial z}\right)^2\right) - \frac{\partial w_x}{\partial y}\frac{\partial w_y}{\partial x} - \frac{\partial w_x}{\partial z}\frac{\partial w_z}{\partial x} - \frac{\partial w_y}{\partial z}\frac{\partial w_z}{\partial y} \tag{4.12}$$

The Q criterion delineates the equilibrium between rotational and strain-induced deformation within a fluid element in the flow field. A positive Q value suggests that rotational effects are predominant, indicative of vortex presence. Utilizing the Q criterion for vortex identification in flow field analysis serves dual purposes: firstly, the magnitude of Q quantifies the vortex intensity, with the extremum of Q signifying the locus of the vortex core. Secondly, the construction of Q iso-surfaces facilitates the assessment of vortex characteristics, including their spatial dimensions and morphology, by analyzing the volume and directional attributes of the regions encapsulated by these surfaces.

Subsequently, we expound upon the application with a compressor example. As portrayed in Fig. 4.7, the illustration of the flow field in the proximity of the tip region of a compressor cascade, under conditions of a Mach number of 0.3 and an incident flow angle of 0°. The Q iso-surface contoured by the pressure can help the identification of vortex structures within the flow field.

The illustration unveils the manifestation of a quintessential "tri-vortex structures," which encompasses the tip separation vortex (TSV) adjacent to the pressure side of the blade, the tip secondary vortex (TSEV) along the suction side, and the tip leakage vortex (TLV) situated on the suction side. A contour plot of the Q criterion on the S3 stream surface at 40% of the blade's chord length, as depicted in Fig. 4.8, enables a distinct differentiation of the vortex structures' intensity and positional relationships within the tip region. The TLV, situated on the suction side of the blade, is the largest in scale but has a comparatively lower vortex intensity when contrasted with the TSV. The TSV, located at the edge of the blade's pressure side, is elongated in the direction of the blade's thickness, with a spatial scale smaller than that of the TLV, yet it possesses the highest vortex intensity. The TSEV, positioned at the edge of the blade's suction side, exhibits the smallest scale and vortex intensity among the three.

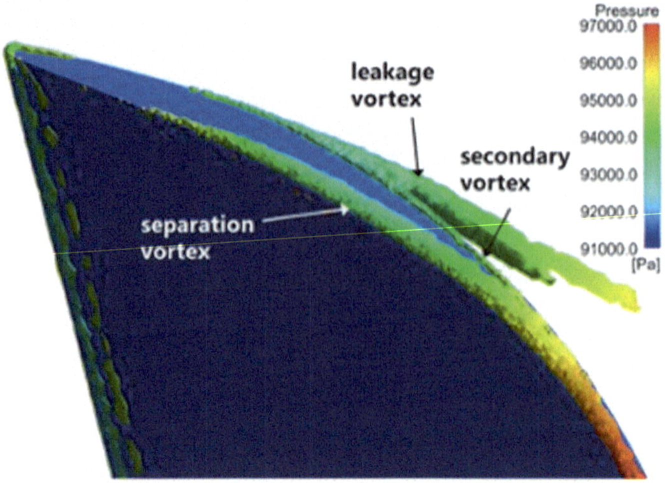

Fig. 4.7 Iso-surface of the Q criterion in the tip region of the compressor cascade ($Q = 2 \times 10^8$)

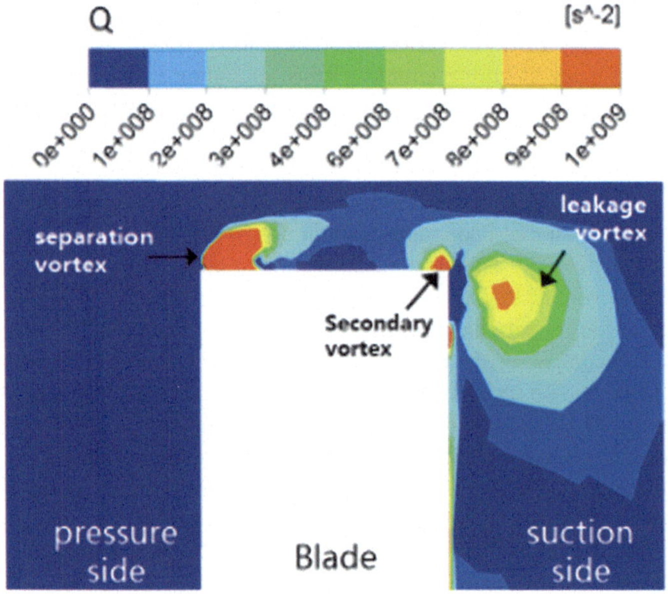

Fig. 4.8 The contour plot at the 40%C section

4.3.2 λ_2 Criterion

The λ_2 criterion was proposed by Jeong and Hussain [14]. Under the assumption of incompressibility and neglecting the unsteady and viscous terms in the Navier–Stokes equations, the following can be derived:

$$\mathbf{A}^2 + \mathbf{B}^2 = -\nabla \cdot (\nabla p)/\rho \tag{4.13}$$

When the symmetric tensor $\mathbf{A}^2 + \mathbf{B}^2$ has two negative eigenvalues, the pressure is at a minimum within the plane spanned by the eigenvectors corresponding to these two negative eigenvalues. When the eigenvalues are ordered as $\lambda_1 > \lambda_2 > \lambda_3$, the existence of two negative eigenvalues implies $\lambda_2 < 0$, hence the method is termed the λ_2 criterion.

In the λ_2 criterion approach, vortices are defined as regions where. However, this method assumes an incompressible fluid and disregards the unsteady and viscous terms, thus making it challenging to accurately identify vortex structures when strong unsteady and viscous effects are present in the flow field. Empirical evidence suggests that for subsonic flow fields, the Q criterion $\lambda_2 < 0$ results closely align with the λ_2 criterion. As illustrated in Fig. 4.9, the corner separation of the blade cascade is depicted using the λ_2 criterion to reveal the separated vortex structures. The λ_2 criterion also visualizes vortex structures using iso-surfaces, which requires the specification of a threshold value; the figure presents results for three different threshold values.

It is evident that the second-generation vortex identification methods, represented by the Q criterion and the λ_2 criterion, are scalar-based and display vortices through the use of isosurfaces. These methods rely on the selection of a threshold value, and variations in the choice of threshold can significantly lead to the depiction of different vortex structures. Additionally, the refinement of flow field analysis using

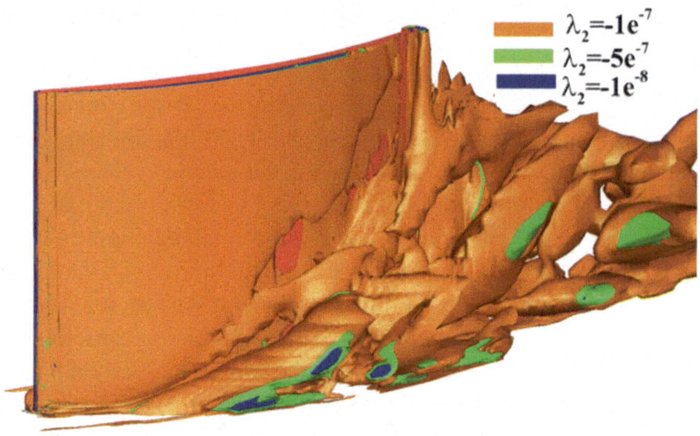

Fig. 4.9 The vortex structure of the compressor corner region based on the λ_2 criterion

DES methods has greatly improved compared to URANS, allowing for the coexistence of weak and strong vortices within the flow field. However, setting the threshold too high may fail to reveal weak vortices, whereas setting it too low can result in an overly blurred representation of the entire field of vortex structures.

4.3.3 Liutex Method

The third-generation vortex identification method known as Liutex was proposed by Professor Chaoqun Liu's team at the University of Texas at Arlington [10]. The aim of this method is to address the reliance on threshold values in the second-generation vortex identification methods and their inability to represent the direction of vortices. The main idea is to use a vector to represent vortices. This vector has a direction that corresponds to the axis of rotation of the vortex, and its magnitude represents the vortex intensity, thereby describing the local fluid rotation using the axis of rotation and its intensity.

Analyzing the velocity gradient tensor $\nabla \cdot \mathbf{W}$, its characteristic equation is given by $|\nabla \cdot \mathbf{W} - \lambda \mathbf{I}| = 0$, which yields three eigenvalues. Since the characteristic equation is a cubic polynomial, there are only two possibilities: either three real eigenvalues λ_1, λ_2, λ_3, or one real eigenvalue and two complex conjugate eigenvalues, denoted as λ_r, $\lambda_{cr} + i\lambda_{cr}$ and $\lambda_{cr} - i\lambda_{cr}$. According to Liu's theorem [9], fluid rotation is represented only when there is one real eigenvalue and two complex conjugate eigenvalues; hence, we focus our study exclusively on this scenario.

The explicit formula for the Liutex vector is given by: $\mathbf{R} = R\mathbf{r}$, where R is twice the magnitude of the local vortex rotational angular velocity, defined by Eq. (4.14), representing the size of the vortex. $\mathbf{r}$ is the real eigenvector of $\nabla \cdot \mathbf{W}$, that is, $(\nabla \cdot \mathbf{W}) \cdot \mathbf{r} = \lambda_r \mathbf{r}$, indicating the direction of the local vortex rotation.

$$R = (\boldsymbol{\omega} \cdot \mathbf{r}) - \sqrt{(\boldsymbol{\omega} \cdot \mathbf{r})^2 - 4\lambda_{ci}^2} \tag{4.14}$$

Like the Q criterion and the λ_2 criterion, the magnitude R of the Liutex is also a scalar that can be represented by iso-surfaces to illustrate vortex structures. However, unlike the second-generation criteria, R represents rigid rotation and is not contaminated by stretching and shearing, being twice the angular velocity of local fluid rotation. With the Liutex method, both strong and weak vortices can be displayed simultaneously, and vector information can also be obtained. Figures 4.10 and 4.11 compare the vortex identification capabilities of the Q criterion and the Liutex method using the same simulation results. From Fig. 4.10, it can be seen that when the iso-surface value of the Q criterion is too large, many vortex structures cannot be displayed (Fig. 4.10a). Continuously reducing the value of Q, as shown in Fig. 4.10c, results in a large iso-surface area that is difficult to discern. In contrast, with Liutex, a single iso-surface can clearly display relatively strong TLV and comparatively weaker

(a) Q =1 × 10^5 (b) Q =1 × 10^4 (c) Q =1 × 10^2

Fig. 4.10 Vortex structure identification in the compressor blade tip clearance based on the Q criterion

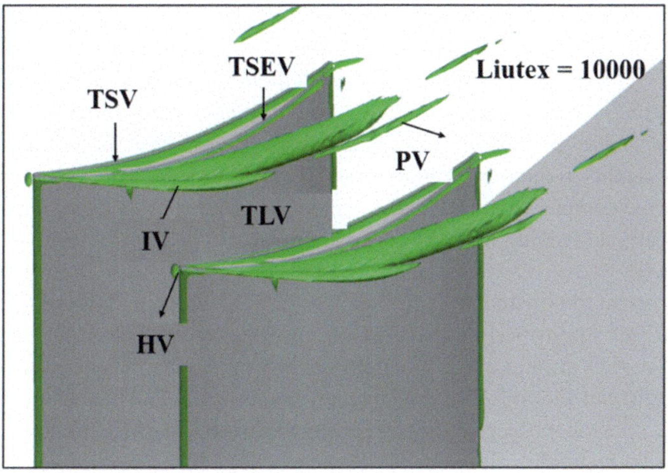

Fig. 4.11 The vortex structure of the compressor tip region based on the Liutex method ($Liutex_{mag} = 1 \times 10^4$)

induced vortex, TSV and TSEV, which aids in the analysis of the mechanisms of interference between vortex structures.

4.4 The Temporal-Spatial Decoupling Method for Highly Coupled Flow Fields

This section leverages advanced data mining techniques, namely POD, DMD and SPOD to analyze high-fidelity flow field data in the corner region of a compressor. By employing these methods, we aim to decouple the intricate spatiotemporal flow field information, thereby facilitating the analytical study of the intricate interferences, inductions, and evolutionary processes within complex turbulent structures. Subsequent discussions will delve into the basic conceptual framework and merits of these methodologies.

4.4.1 POD and Its Application in the Flow Field Analysis

POD is a method derived from statistical analysis of vector data. Since Lumley (1967) introduced this method into turbulence research [15], POD has played a significant role in the processing and analysis of data in fluid mechanics. It stands as one of the essential tools for analyzing turbulent structures and the underlying mechanisms within flow fields. When applied to flow field analysis, the POD method sorts the entire flow field fluctuations based on their energy magnitudes, enabling the identification of turbulent structures that contribute the most to the unsteady fluctuation energy and the spatiotemporal characteristics of the flow field.

However, as the spatial resolution of flow field experiments or numerical simulations has improved, the amount of information in flow fields has substantially increased. The immense dimensions of the resulting matrices make Lumley's proposed POD method (also known as direct POD method) challenging to solve. Consequently, Sirovich, leveraging the temporal dimension of the data being much less than the spatial dimension, employed spatiotemporal transformation to address the dimensionality reduction of high-dimensional data [16]. This approach significantly lowered computational costs and became known as the snapshot POD method, thereby greatly advancing research on coherent structures of turbulence in complex flow fields.

It is important to note that both the snapshot POD method and the spectral Proper Orthogonal Decomposition method, which will be discussed later, are abbreviated as SPOD. Readers should exercise caution in distinguishing between the two when reviewing relevant literature. In this chapter, the term SPOD is used specifically to refer to the spectral Proper Orthogonal Decomposition method, in contrast to the snapshot POD method.

Drawing from Chen's [17] research, this section delineates the fundamental yet pivotal Snapshot POD method. Emphasis is placed on elucidating its conceptual framework, procedural construction, and physical implications.

The essence of PDF method is to achieve an optimal low-dimensional approximation of a given set of high-dimensional data. The core idea is to decompose the original flow field into an algebraic sum of a function dependent only on time and a set of standard basis functions dependent only on space. The key to this process is to derive a series of standard basis functions (often referred to as "modes") from known experimental or numerical simulation data, ensuring that these basis functions are the optimal orthogonal basis functions in the least-squares sense. In other words, for a matrix ($\mathbf{G}_N^K = \{g_1, g_2, \ldots, g_n, \ldots, g_N\}$) composed of scalar data from N snapshots at K data points (i.e., spatial grid points) at each time step, with g_n representing the flow field at the nth time step, as shown in Fig. 4.12, the goal is to find a set of basis functions $\boldsymbol{\Psi}_N^K = \{\psi_1, \psi_2, \ldots, \psi_N\}^T$ such that:

$$\max\left\{\frac{1}{N}\sum_{i=1}^{N}\left|\left(g_i, \psi_i^T\right)\right|^2\right\} \tag{4.15}$$

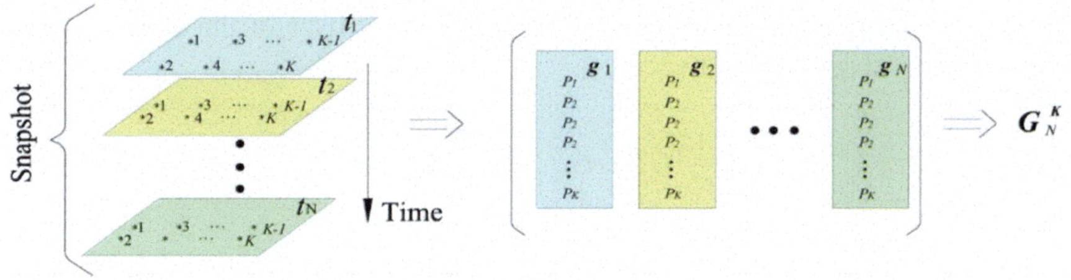

Fig. 4.12 The input of the POD, N represents the snapshot number and K represents the grid number

$$s.t. \ \left(\left(\boldsymbol{\Psi}_N^K \right)^T, \boldsymbol{\Psi}_N^K \right) = \mathbf{I}_N^N, \tag{4.16}$$

where, |·| represents the norm, (·, ·) represents the inner product, and $\boldsymbol{I}$ represents the identity matrix.

How can we construct the standard orthogonal basis functions, that is, $\boldsymbol{\Psi}_N^K$? Given that our focus is on the fluctuations in the flow field, we consider the unsteady fluctuation data matrix, $\boldsymbol{G'}_N^K = \{g'_1, g'_2, \ldots, g'_n, \ldots, g'_N\}$, which represents the difference between the original flow field $\mathbf{G}_N^K$ and its time-averaged value. The fluctuation flow field at time n is given by:

$$g'_n = g_n - \frac{1}{N}\sum_{i=1}^{N} g_i, \quad n \in [1, N]. \tag{4.17}$$

Thus, the fluctuation flow fields at N time steps can be represented in matrix form as:

$$\boldsymbol{G'}_N^K = \left[g'_1 \ g'_2 \cdots g'_N \right], \tag{4.18}$$

where the nth column represents the fluctuation flow field at time n.

In POD analysis, the primary concern is the number of spatial discrete points, the number of snapshots, and the magnitude of the corresponding physical quantities, without distinguishing between spatial locations. Therefore, during the analysis, the three-dimensional spatial data at a given time are sequentially arranged in a one-dimensional format according to the computational coordinate order, as illustrated in Fig. 4.12.

Use the Lagrange multiplier method to solve the optimization problem as Eqs. (4.15) and (4.16) shows. The Lagrange equation is

$$J(\boldsymbol{\Psi}) = \sum_{j=1}^{N_m} \frac{1}{N} \left| (\boldsymbol{g}_i, \boldsymbol{\Psi}) \right|^2 - \lambda\left(\left(\boldsymbol{\Psi}^{\mathrm{T}}, \boldsymbol{\Psi} \right) - \boldsymbol{I} \right), \tag{4.19}$$

where, λ is the Lagrange multiplier. Take partial derivatives of the equation with respect to $\boldsymbol{\Psi}$ and set it equal to zero:

$$\left(\frac{1}{N}\left(\boldsymbol{G}_N^{\prime K}\right)^T \boldsymbol{G}_N^{\prime K} - \boldsymbol{\Lambda}_{\mathrm{N}}^{\mathrm{N}}\right)\boldsymbol{A}_N^N = 0 \tag{4.20}$$

Then define the correlation matrix $\boldsymbol{C}_N^N$ representing the temporal correlation between any two snapshots of the flow field as

$$\boldsymbol{C}_N^N = \frac{1}{N}\left(\boldsymbol{G}_N^{\prime K}\right)^T \boldsymbol{G}_N^{\prime K} \tag{4.21}$$

and Eq. (4.20) can be represented as

$$\boldsymbol{C}_N^N \boldsymbol{A}_N^N = \boldsymbol{A}_N^N \boldsymbol{\Lambda}_{\mathrm{N}}^{\mathrm{N}} \tag{4.22}$$

where, $\boldsymbol{\Lambda}_N^N = diag(\lambda_1, \lambda_2, \ldots, \lambda_N)$ and $\lambda_1 > \lambda_2 > \cdots > \lambda_N \geq 0$.

The eigenvalues represent the turbulent energy corresponding to POD modes, and the total turbulent energy E_{all} of the turbulent flow field is

$$E_{all} = \sum_{i=1}^{N} \lambda_i \tag{4.23}$$

The POD modes $\boldsymbol{\Psi}_N^N$ and the eigenvectors $\boldsymbol{A}_N^N$ have the following relationship

$$\boldsymbol{\Psi}_N^{\prime N} = \boldsymbol{G}_N^{\prime K} \boldsymbol{A}_N^N \tag{4.24}$$

$$\boldsymbol{\Psi}_N^K = \left[\frac{\psi_1'}{\sqrt{\lambda_1}}, \frac{\psi_2'}{\sqrt{\lambda_2}}, \cdots, \frac{\psi_n'}{\sqrt{\lambda_n}}\right] \tag{4.25}$$

The original flow field $G_N^{\prime K}$ can be decomposed into the temporal coefficients $\boldsymbol{B}_N^N$ and spatial POD modes $\boldsymbol{\Psi}_N^K$, namely

$$\boldsymbol{G}_N^{\prime K} = \left(\boldsymbol{B}_N^N\left(\boldsymbol{\Psi}_N^K\right)^T\right)^T \tag{4.26}$$

where, $\boldsymbol{B}_N^N = \left(\boldsymbol{G}_N^{\prime K}\right)\boldsymbol{\Psi}_N^K$ is the temporal coefficient matrix. The column of $\boldsymbol{B}_N^N$ represents the unsteady pulsation patterns of the corresponding POD mode, which can be used to obtain the frequency components of the POD mode, and the line of $\boldsymbol{B}_N^N$ represents the contribution of the POD mode.

Readers are encouraged to delve into supplementary literature [18] to explore alternative derivational approaches, which enriches their comprehension. It is noteworthy that despite varied derivational approaches, the core tenets remain congruent.

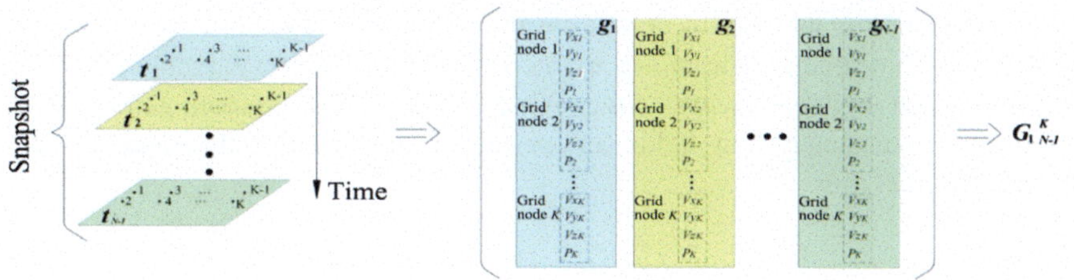

Fig. 4.13 The input of the DMD, N represents the snapshot number and K represents the grid number

4.4.2 DMD and Its Application in the Flow Field Analysis

DMD is a novel method for extracting spatiotemporal features embedded in a flow. It was first proposed by Schmid and Sesterhenn in 2008 at an academic conference [20], and quickly found applications in the analysis of fluid dynamics data [21–23]. The DMD method divides the flow field into multiple modes based on frequency, with each mode describing a single-frequency flow structure. Additionally, it can characterize the stability of each mode, thereby uncovering flow structures that induce instability in the entire flow field. This section will introduce the standard DMD method, focusing on its principles, application methodologies, and the frequency, growth rate, and individual flow field reconstruction techniques utilized in flow field analysis.

For a given time series data, DMD identifies it as a series of dynamic units described by mode, frequency, and growth rate, namely $\{\boldsymbol{\Phi}, f, s\}$. Each mode identified by DMD exhibits two characteristics: a fixed frequency and a growth rate. Each mode can reconstruct a flow field with a single pulsation frequency. The growth rate represents the growth or decay of unsteady pulsation amplitudes and is used to characterize the stability of the mode.

The input data for DMD consists of a series of snapshots at N time instances, denoted as $\{g_1, g_2, \ldots, g_n, \ldots g_{N-1}, g_N\}$, where the time intervals between adjacent snapshots are the same as illustrated in Fig. 4.13. Each snapshot contains the physical field data (such as pressure, velocity components, etc.) at K grid points.[1] We construct the matrix $\boldsymbol{G}_{1\,N-1}^{K}$ from snapshots at time instances 1 to $N-1$ and the matrix $\boldsymbol{G}_{2\,N}^{K}$ from snapshots at time instances 2 to N.

$$\mathbf{G}_{1N-1}^{K} = \{g_1, g_2, \ldots, g_n, \ldots, g_{N-1}\} \tag{4.27}$$

$$\mathbf{G}_{2N-1}^{K} = \{g_2, g_3, \ldots, g_n, \ldots, g_N\} \tag{4.28}$$

[1] When G contains one physical quantity, it refers to the *standard DMD*, and when G contains multiple physical quantities, it refers to the *composite DMD*.

DMD inherently address linear time-invariant dynamical systems,[2] where a snapshot at a certain time is determined by the snapshot at the previous time, and this transformation is independent of time. When applying DMD to data generated by nonlinear and unsteady flow fields, an operator A approximates the tangent to the flow field, satisfying:

$$g_{n+1} = Ag_n \quad n = [1, 2, \ldots, N-1] \tag{4.29}$$

where, A is the transfer function to link the g_n and g_{n+1}.

The matrix $\mathbf{G}_1{}_{N-1}^{K}$ and $\mathbf{G}_2{}_{N-1}^{K}$ can be expressed as

$$\mathbf{G}_1{}_{N-1}^{K} = \mathbf{A}_K^K \cdot \mathbf{G}_2{}_{N-1}^{K} \tag{4.30}$$

Perform singular value decomposition (SVD) on $\mathbf{G}_1{}_{N-1}^{K}$ yields

$$\mathbf{G}_1{}_{N-1}^{K} = \mathbf{U}_r^K \cdot \boldsymbol{\Sigma}_r^r \cdot \left(\mathbf{V}_r^{N-1}\right)^T \tag{4.31}$$

where, r is the rank of the $\mathbf{G}_1{}_{N-1}^{K}$.

For $\boldsymbol{A}$ is determined by the control equations, this subsection use the similar matrices method based on SVD to construct low dimensional matrix $\widetilde{\mathbf{A}}_r^r$ to approximate the high dimensional matrix $\mathbf{A}_K^K$, to get rid of the control equations and implement DMD in a purely data-driven manner.

$$\mathbf{A}_K^K = \mathbf{U}_r^K \cdot \widetilde{\mathbf{A}}_r^r \cdot \left(\mathbf{U}_r^K\right)^T \tag{4.32}$$

where, $\left(\mathbf{U}_r^K\right)^T \cdot \mathbf{U}_r^K = \mathbf{I}_r^r$ and $\boldsymbol{I}_r^r$ is the identity matrix. The calculation of $\tilde{\boldsymbol{A}}_r^r$ satisfy the minimization problem

$$\underset{\widetilde{\mathbf{A}}}{\text{minmize}} \left\| \mathbf{G}_2{}_{N-1}^{K} - \mathbf{U}_r^K \cdot \widetilde{\mathbf{A}}_r^r \cdot \boldsymbol{\Sigma}_r^r \cdot \left(\mathbf{V}_r^{N-1}\right)^T \right\|_F^2, \tag{4.33}$$

where, $\|\cdot\|_F$ represents the Forbenius norm. Therefore, the low dimensional matrix $\tilde{\boldsymbol{A}}$ can be expressed as

$$\widetilde{\mathbf{A}} = \left(\mathbf{U}_r^K\right)^T \cdot \mathbf{G}_2{}_{N-1}^{K} \cdot \mathbf{V}_r^{N-1} \cdot \left(\boldsymbol{\Sigma}_r^r\right)^{-1}, \tag{4.34}$$

where, $\tilde{\boldsymbol{A}}$ is the similarity matrix of $\boldsymbol{A}$, which means they have the same eigenvalues reflecting the same dynamic characteristics.

The frequency f and growth rate s of the DMD mode $\boldsymbol{\Phi}$ can be obtained from the eigenvalues of the $\tilde{\mathbf{A}}$. For instance, for the ith eigenvalue λ_i, the frequency f_i and growth rate s_i are respectively given by Eqs. (4.35) and (4.36). The growth rate

[2] The assumption determines the inherently sine instability, i.e., $\frac{dg}{dt} = Bg$. (Note: B is related to A but they're different from each other.)

s_i serves as an indicator of the stability of the ith mode Φ_i: a positive growth rate signifies that the unsteady pulsation of the mode continually increases, indicating instability; conversely, when the growth rate is zero or negative, the mode maintains periodic flow or gradually diminishes in amplitude, suggesting stability.

$$f_i = \mathrm{Im}(\lg \lambda_i)/\Delta t \tag{4.35}$$

$$s_i = \mathrm{Re}(\lg \lambda_i)/\Delta t \tag{4.36}$$

The ith mode Φ_i containing the specific frequency can be obtained through the eigenvectors φ_i of the $\tilde{\boldsymbol{A}}_r^r$.

$$\Phi_i = \mathbf{G}_1{}_{N-1}^K \varphi_i \tag{4.37}$$

where, φ_i is the ith eigenvector, Φ_i is the reconstructed flow field corresponding to a specific frequency in a certain moment. The L_2 norm of the ith mode Φ_i is commonly utilized to assist in selecting modes that constitute the primary structures of the flow field. Typically, modes with larger values of $\|\Phi_i\|_2$ are chosen as the primary modes for analysis.

However, it's important to note that when applying the DMD, considerations of numerical accuracy and physical interpretability are necessary. The former is influenced by factors such as parameter settings, noise, and methods for selecting dominant modes, leading to various DMD variants. Regarding the latter, the standard DMD described above is essentially applicable to linear time-invariant flow fields. Although Schmid pointed out the applicability of DMD to nonlinear flow fields in 2008, the physical interpretation of the dynamical features identified directly through the standard DMD algorithm is ambiguous and even meaningless. Therefore, readers should perform correlation analysis on the results to 'synthesize' flow field structures with practical physical meanings, or apply Koopman analysis or Extended DMD (EDMD) to achieve feature identification in nonlinear systems.

In fact, over the 15 years of development since its inception, the DMD has been advancing towards accuracy, robustness, and efficiency. Readers are highly recommended the Schmid's outstanding review [24] on the DMD algorithm, which will help readers better grasp the development trajectory of the DMD algorithm and stay updated on the latest research trends.

4.4.3 SPOD Method

Compared with the traditional proper orthogonal decomposition (POD) method, SPOD can extract the physical modes of a specific frequency band in an energy mode while obtaining all the modes arranged in descending order of energy, taking into account the frequency domain information of the flow field. Compared with the dynamic mode decomposition (DMD) method, the frequency mode extracted by this method is more representative [19]. Finally, the principle of the SPOD method is as follows:

A snapshot $Q_{N_t}^K = \left[h_1', h_2', \ldots, h_{N_t}'\right]$ based on fluctuation vectors is constructed first, where N_t is the total number of snapshots and K is the total number of grid cells. Using the Welch periodogram method, the data matrix $\mathbf{Q}_{N_t}^K$ is then segmented into N_b sub-block matrixes, each sub-block matrix comprising N_f snapshots. Equation (4.38) shows the n_{th} sub-block matrix ${Q^{(n)}}_{N_f}^K$:

$$Q_{N_f}^{(n)K} = \left[h_1'^{(n)}, h_2'^{(n)}, \ldots, h_{N_f}'^{(n)}\right];\ 1 \le n \le N_b \tag{4.38}$$

where $h'^{(n)}_k$ represents the k_{th} snapshot in the n_{th} sub-block matrix. Since each block is assumed to be a statistically independent realization under the ergodicity hypothesis, we subsequently performed a discrete Fourier transform on each block to transform the data into the frequency domain; then, k_{th} snapshot of the n_{th} sub-block is as follows:

$$\hat{\mathbf{h}}_k^{(n)} = \frac{1}{\sqrt{N_f}} \sum_{j=1}^{N_f} w_j \mathbf{h}_j^{(n)} \mathrm{e}^{-\mathrm{i}2\pi(k-1)[(j-1)/N_f]} \tag{4.39}$$

The weights w_j represent the nodal value of a window function that can be used to reduce spectral leakage caused by the non-periodicity of the snapshot data in each frequency-domain block [25].

Then, we reshape the frequency domain matrices according to their frequency. As shown in Eq. (4.40), $\widehat{\mathbf{Q}}_k$ is constructed by all the snapshots $\hat{\mathbf{h}}_k^{(n)}$ whose frequencies are k. The weighted cross spectral density (CSD) matrix for the k_{th} element can be designed by Eq. (4.41):

$$\hat{\mathbf{Q}}_k = \left[\hat{\mathbf{h}}_{\mathbf{k}}^{\mathbf{1}}, \hat{\mathbf{h}}_{\mathbf{k}}^{\mathbf{2}}, \ldots, \hat{\mathbf{h}}_{\mathbf{k}}^{\mathbf{N_b}}\right];\ 1 \le n \le N_b \tag{4.40}$$

$$\mathbf{S}_k = \frac{1}{N_b} \mathbf{W}^{\frac{1}{2}} \hat{\mathbf{Q}}_k^* \hat{\mathbf{Q}}_k \mathbf{W}^{*\frac{1}{2}} \tag{4.41}$$

Furthermore, the eigenvalue decomposition of the CSD matrix $\mathbf{S}_k$ is carried out. The mode $\boldsymbol{\Phi}_k$ corresponding to the frequency k is obtained according to the eigenvalue matrix $\boldsymbol{\Lambda}_k$ and the eigenvector matrix $\boldsymbol{\Psi}_k$, as shown in Eq. (4.42). The energy of each mode $\phi_k^{(n)}$ is characterized by the component $\boldsymbol{\lambda}_k^{(n)}$ of the eigenvalue $\boldsymbol{\Lambda}_k$.

$$\boldsymbol{\Phi}_k = \hat{\mathbf{Q}}_k \boldsymbol{\Psi}_k \boldsymbol{\Lambda}_k^{(-1/2)} \tag{4.42}$$

In this study, the flow field reconstruction is based on the time domain reconstruction method [26]. After the original snapshot data $\mathbf{Q}_{N_t}^K$ expressed as a linear combination of SPOD modes (see Eq. 4.43), the coefficient conversion matrix $\widetilde{\mathbf{A}}$ is solved as shown in Eq. (4.44). The weight matrix $\mathbf{W}$ changes with the parameters selected in the snapshot.

$$\mathbf{Q}_{N_t}^{K} \approx \widetilde{\mathbf{Q}}\widetilde{\mathbf{A}} \tag{4.43}$$

$$\widetilde{\mathbf{A}} = \left(\widetilde{\boldsymbol{\Phi}}^{*}\mathbf{W}\,\widetilde{\boldsymbol{\Phi}}\right)^{-1}\widetilde{\boldsymbol{\Phi}}^{*}\mathbf{W}\mathbf{Q}_{N_t}^{K} \tag{4.44}$$

The matrix $\widetilde{\boldsymbol{\Phi}}$ contains the SPOD modes of all frequencies, namely:

$$\widetilde{\Phi} = \left[\phi_1^1\phi_2^1, \phi_2^2, \ldots, \phi_2^{N_b}, \ldots, \phi_{N_f}^1, \phi_{N_f}^2, \ldots, \phi_{N_f}^{N_b}\right] \tag{4.45}$$

Finally, the fluctuating flow field ($\widetilde{h'_m}$) at the m_{th} moment can be reconstructed by Eq. (4.46).

$$\widetilde{h'_m} = \left(\widetilde{\boldsymbol{\Phi}}^{*}\mathbf{W}\,\widetilde{\boldsymbol{\Phi}}\right)^{-1}\widetilde{\boldsymbol{\Phi}}^{*}\mathbf{W}h'_m \tag{4.46}$$

4.4.4 Analysis of POD and DMD

(1) **Setup of a cylinder flow simulation example**

This subsection uses the cylinder flow simulation to verify the accuracy of the time–space decoupling method. The computational domain, as depicted in Fig. 4.14, encompasses a rectangular region with dimensions of $25D \times 12.5D$. Within this domain, a cylinder with a diameter of $D = 4$ mm is positioned, bounded by upper and lower walls. The inflow velocity is set at 0.05 m/s, utilizing water as the working fluid with a dynamic viscosity of $\mu = 1.7894 \times 10^{-5}$ kg s/m^2. The Reynolds number is maintained at $Re = 137$. For unsteady computations, a physical time step of 0.1 s is employed, utilizing a dual-time-stepping method with 50 iterations. The computational framework utilizes *Fluent* software, with the $k - \varepsilon$ turbulence model selected for analysis.

Following the unsteady computation, the flow field exhibits periodic fluctuations, achieving unsteady convergence after 31 s of physical time. Subsequently, data results are saved at intervals of 0.1 s for a total of 690 time steps. To address the challenge of large data volume, the study focuses on discernible fluctuations within specific regions of interest (ROI) within the flow field. In this instance, the square region delineated by the black box in Fig. 4.15, characterized by $x \in (0.3, 0.6)$ and $y \in (0.15, 0.35)$, is selected for analysis. This region's pressure field data is subjected to POD and DMD for spatiotemporal decoupling. Furthermore, pressure variations

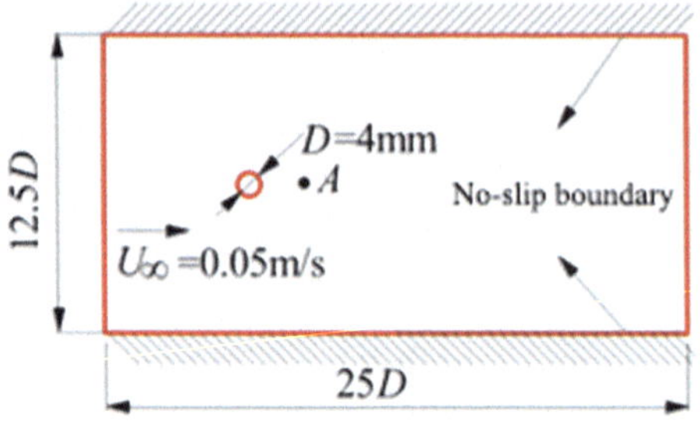

Fig. 4.14 The diagram of computational domain for cylinder flow

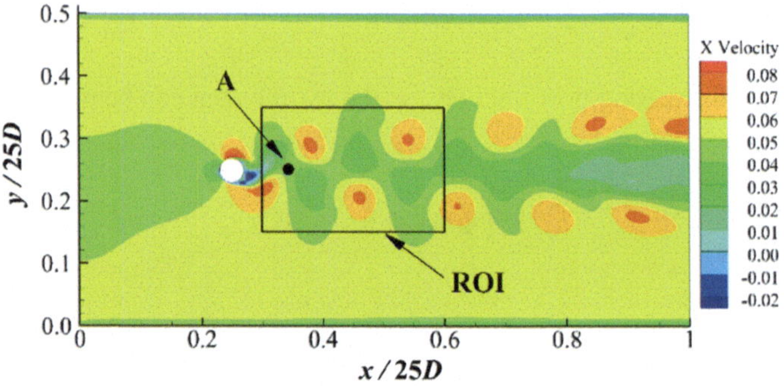

Fig. 4.15 The region of interests and monitoring point A

over time at monitoring point A, located 10 mm downstream of the cylinder's rear, as depicted in Fig. 4.14, are recorded to serve as a calibration baseline.

(2) **Verification of time–space decoupling method**

The time-domain results of the unsteady pressure at monitoring point A are depicted in Fig. 4.16, exhibiting periodic fluctuations in pressure characterized by two distinct fixed frequencies of varying amplitudes. The frequency distribution is presented in Fig. 4.17, wherein the primary pulsation frequency is 0.576 Hz, the secondary pulsation frequency is 0.293 Hz, and 1.1632 Hz represents the second harmonic of 0.576 Hz.

The pulsatile pressure field within the ROI undergoes POD transformation. The modal-energy results in Fig. 4.18 indicate that the first four modes encapsulate the predominant energy of the flow field. Given that POD analysis is conducted on the fluctuating component of the flow field, i.e., the difference between the original and time-averaged flow fields, the first four modes exhibiting higher energy, signify the principal energy of the unsteady fluctuations. Consequently, the modal coefficients characterizing the frequency and phase information of Mode 1 to Mode 4 are extracted for FFT analysis, as delineated in Fig. 4.19. It is discernible that Mode 1 (Fig. 4.19a) and Mode 2 (Fig. 4.19b) manifest two frequencies: 0.293 and 0.576 Hz,

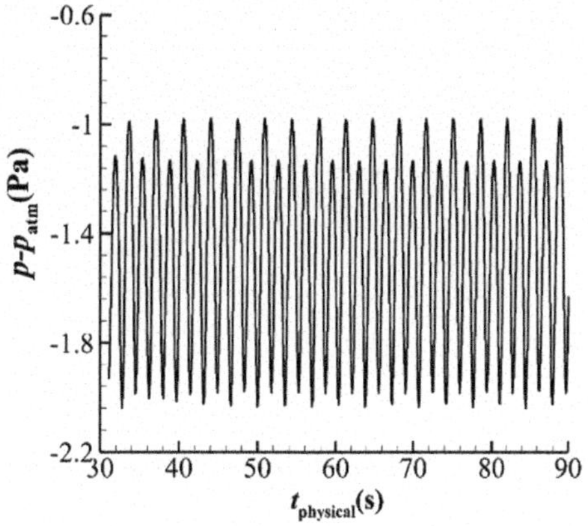

Fig. 4.16 The time-domain graph of pressure pulsations of monitoring A

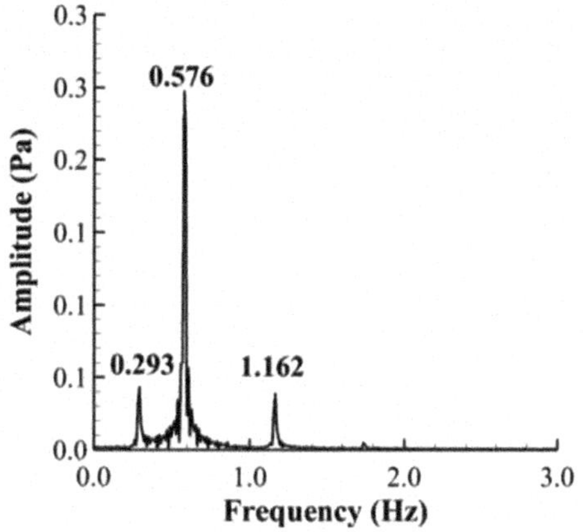

Fig. 4.17 The FFT transformed spectrum graph of pressure pulsations

with 0.293 Hz predominating as the principal pulsation frequency. Analogously, in Mode 3 (Fig. 4.19c), 0.293 and 0.576 Hz are present, albeit with 0.576 Hz emerging as the primary pulsation frequency. In Mode 4 (Fig. 4.19d), solely the pressure pulsation frequency at 0.576 Hz is evident. Overall, the POD analysis delineates that the primary pulsation frequencies within the flow field coincide with 0.293 and 0.576 Hz, consistent with the findings at monitoring point A. However, the divergence lies in the higher amplitude frequency at 0.576 Hz for monitoring point A, while for the POD analysis, the primary pulsation frequency is 0.293 Hz in Mode 1 and Mode 2, where the pulsation energy is more pronounced. This discrepancy arises due to the point-specific nature of FFT analysis, whereas POD analysis encapsulates the primary pulsations within a spatial region, rather than individual points. This underscores the advantage of POD analysis, as it captures more representative flow field characteristics.

For the pressure field within the ROI, the DMD transformation reveals the relationship between the L_2 and frequency of the modes in Fig. 4.20. Each point on the

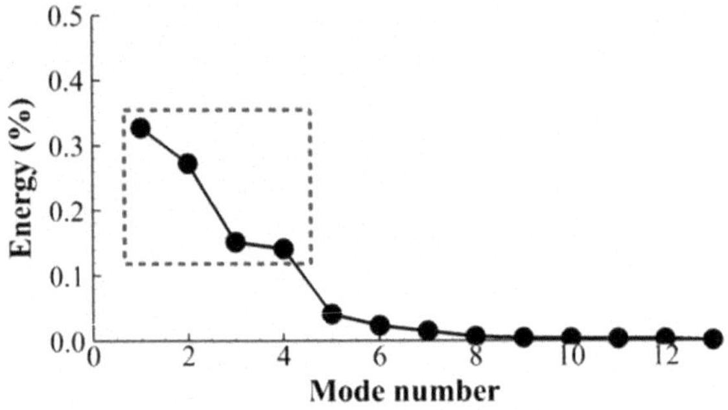

Fig. 4.18 POD mode energy percentage

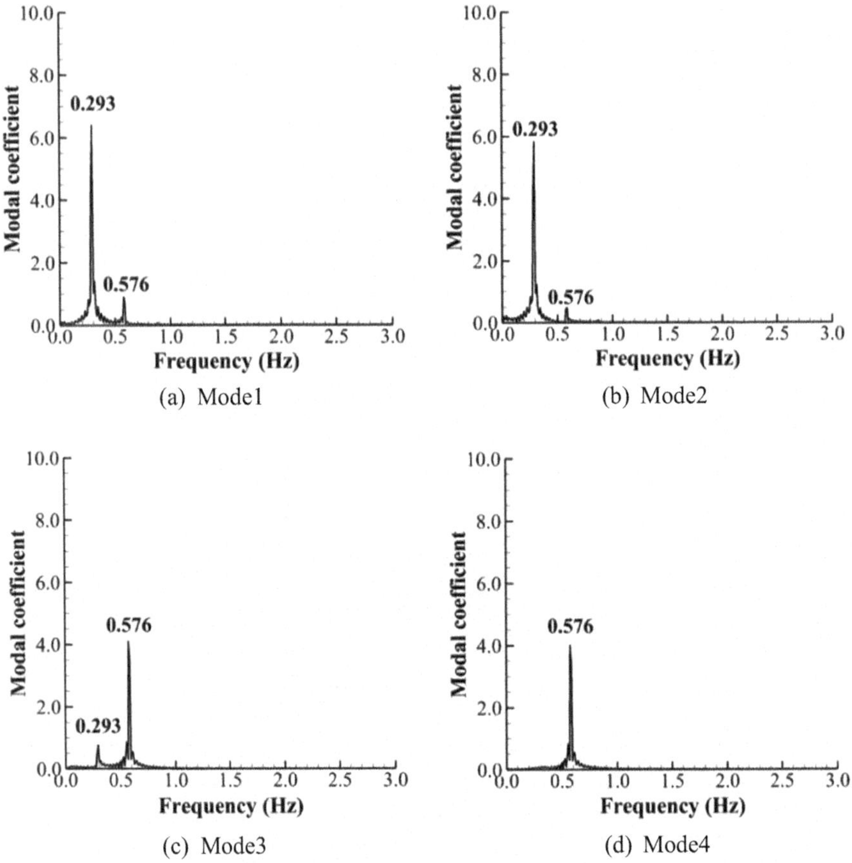

Fig. 4.19 The FFT results of the coefficients of the first four POD modes

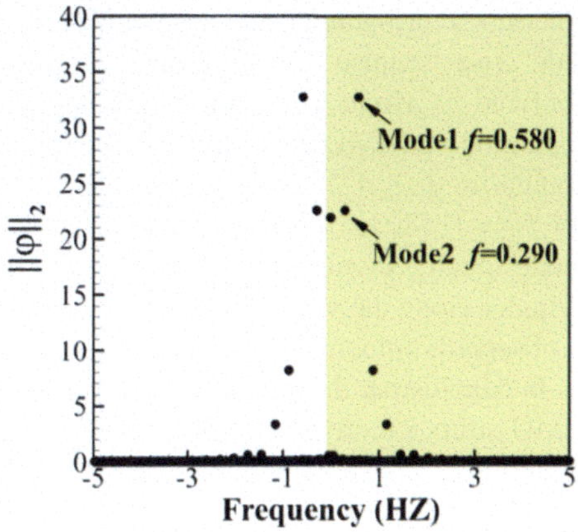

Fig. 4.20 The relationship between the L_2 norm of the DMD modes and the frequency

graph represents a mode, and the symmetry about $f = 0$ Hz is a consequence of the frequency values being the imaginary parts of the eigenvalues (Eq. 4.35), with the negative part representing conjugate poles. Therefore, the analysis focuses solely on modes for which $f \geq 0$, as delineated by the yellow region in Fig. 4.20. Modes with $f = 0$ represent the time-averaged flow field which are disregarded in this discussion. It is generally understood that higher L_2 values of the modes signify a greater impact on the flow field. Consequently, the first two modes are selected based on their higher L_2 values, with $f = 0.580$ Hz and $f = 0.290$ Hz. Table 4.1 provides a comparison between the results of POD, DMD, and the FFT results at monitoring point A. It is evident that while POD accurately captures the energy of the main pulsations, there is a slight discrepancy in the DMD results (approximately 1%).

(3) Comparative analysis of POD and DMD

Comparison between POD and DMD: Figures 4.21 and 4.22 illustrate the reconstructed transient pressure fields of Mode 1 and Mode 3 obtained from POD for a specific moment. In Mode 1 the dominant pulsation with $f = 0.293$ Hz is prominent, whereas in Mode 3 the dominant pulsation with $f = 0.576$ Hz is prevalent. On the other hand, Figs. 4.23 and 4.24 present the results from DMD, corresponding to the transient pressure field reconstructions for frequencies $f = 0.290$ Hz and $f = 0.580$ Hz. Comparing Fig. 4.21 (POD, result for $f = 0.293$ Hz) with Fig. 4.23 (DMD, result for $f = 0.290$ Hz), the qualitative analysis reveals that the isobar

Table 4.1 The comparison of the flow field frequency obtained by different data analysis methods

	f_1 (Hz)	f_2 (Hz)	Error (f_1) (%)	Error (f_2) (%)
FFT	0.576	0.293	–	–
POD	0.576	0.293	0.00	0.00
DMD	0.580	0.290	0.69	1.02

contours of pressure pulsations exhibit general consistency. Discrepancies in magnitude arise because, in POD, pressure pulsations at this frequency are not solely confined to Mode 1 but are distributed across other modes as well. Conversely, each mode in DMD encapsulates all components of a single frequency. In contrast, comparing Fig. 4.22 (POD, result for $f = 0.576\,\text{Hz}$) with Fig. 4.24 (DMD, result for $f = 0.580\,\text{Hz}$), the center of the pressure isobar contours for this frequency mode ideally should align at $y = 0.25\,\text{m}$ (i.e., coinciding with the center of the cylinder along the y-axis). However, in the POD result (Fig. 4.22), a slight deviation is observed, influenced by the flow field at $f = 0.293\,\text{Hz}$.

In conclusion, the frequency analysis of flow fields conducted through POD and DMD offers a more comprehensive understanding of regional pulsation characteristics compared to conventional single-point FFT analysis. POD excels in extracting

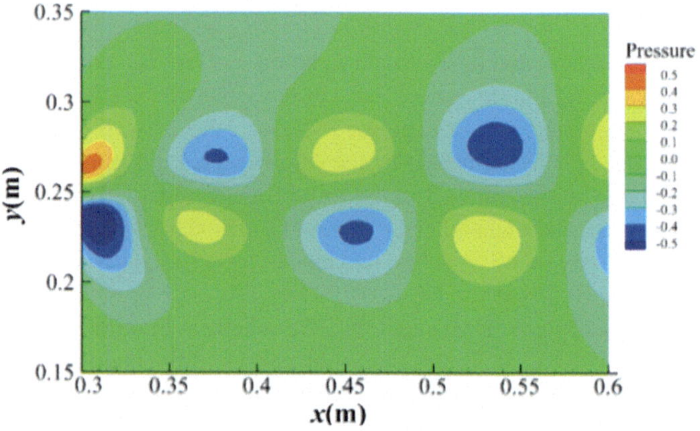

Fig. 4.21 POD $f = 0.293\,\text{Hz}$ dominates the main pulsation of the modal flow field (Mode 1)

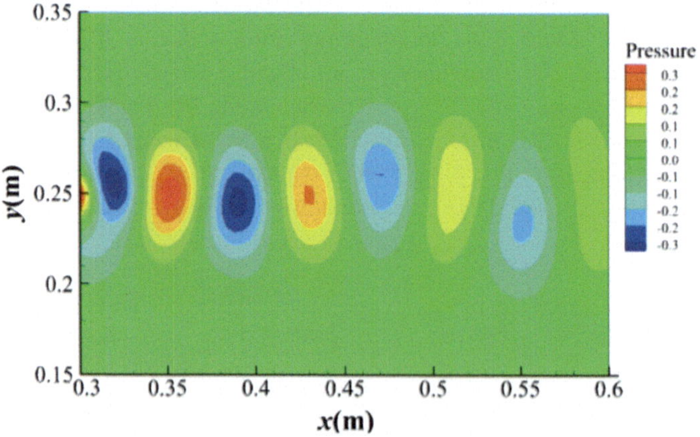

Fig. 4.22 POD $f = 0.576\,\text{Hz}$ dominates the main pulsation of the modal flow field (Mode 3)

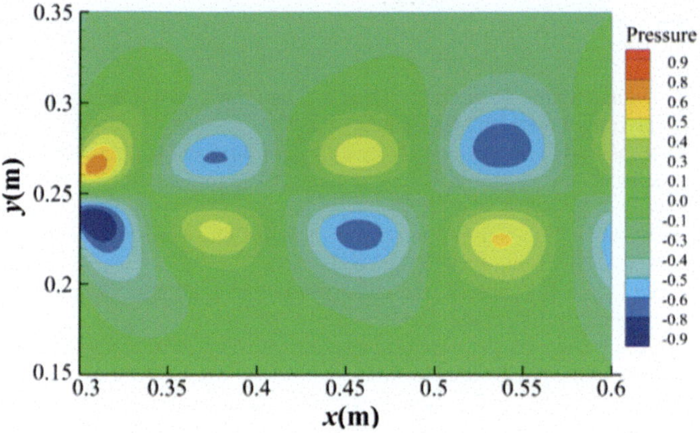

Fig. 4.23 DMD modal flow field with pulsation frequency $f = 0.290$ Hz

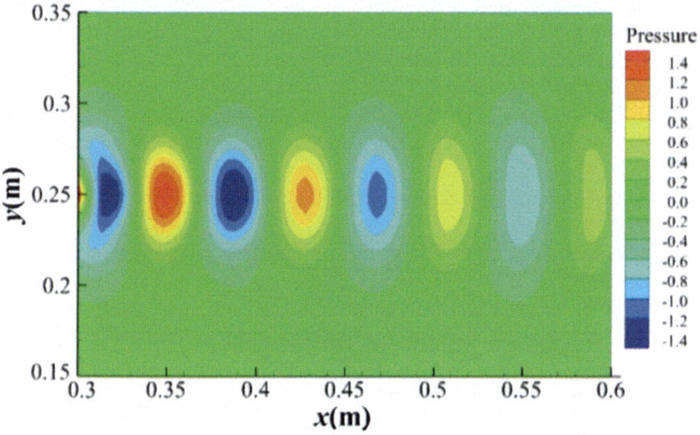

Fig. 4.24 DMD modal flow field with pulsation frequency $f = 0.580$ Hz

flow structures with elevated pulsation energy. However, its limitation arises from the amalgamation of primary frequencies within a singular mode, thereby hindering the discernment of distinct pulsation sources at varying frequencies. Conversely, DMD presents a favorable approach by segregating the flow field into distinct frequency components. Each mode in DMD has a single frequency, facilitating a clearer delineation of the pulsation sources, especially in scenarios where the flow field exhibits multiple frequencies with significant mutual coupling.

References

1. Hu, G S. Digital Signal Processing: Theory, Algorithms, and Implementation. Beijing: Tsinghua University Press, 2012 (in Chinese).
2. Gao L, Li R, Miao F, Cai Y. Unsteady investigation on tip flow field and rotating stall in counter-rotating axial compressor. Journal of Engineering for Gas Turbines and Power, 2015, 137(7): 1–11.
3. Levy Y, Pismenny J. The Number and Speed of Stall Cells During Rotating Stall. Atlanta: Proceedings of ASME Turbo Expo 2003, 2003.
4. Wang C, Yue L. Identification of stall characteristics of axial flow compressor. Journal of Aerospace Power, 2011(08): 1887–1892.
5. Luo Z, Li J. Experiment on single axial compressor of characteristic of rotating stall. Journal of Aerospace Power, 2013, 28(2): 426–431.
6. Helmholtz H. Über Integrale der hydrodynamischen Gleichungen, welche den Wirbelbewegungen entsprechen. Journal für die reine und angewandte Mathematik (Crelles Journal), 1858 (55): 25–55.
7. Robinson S K. Coherent Motions in the Turbulent Boundary Layer. Annual Review of Fluid Mechanics, 1991, 23(1): 601–639.
8. Wang Y, Yang Y, Yang G, et al. DNS Study on Vortex and Vorticity in Late Boundary Layer Transition. Communications in Computational Physics, 2017, 22(02): 441–459.
9. Liu C. Liutex-third generation of vortex definition and identification methods. Acta Aerodynamica Sinica, 2020, 38(3): 413–431.
10. Liu Chaoqun, Gao Yisheng, Tian Shuling, Dong, Xiangrui. Rortex—A new vortex vector definition and vorticity tensor and vector decompositions. Physics of Fluids, 2018, 30(3): 035103.
11. Hunt J C R. Eddies Stream, and Convergence Zones in Turbulent Flows. Center for Turbulence Research. N88-24538, 1988.
12. Eugene de Villiers. The Potential of Large Eddy Simulation for the Modeling of Wall Bounded Flows. London: Imperial College of Science, 2005.
13. Zhang J, Deng X Y. Vortex Identification Criteria and Their Application in PIV Technology. Yanji: Conference on Modern Experimental Aerodynamics, 2009.
14. Jeong Jinhee, Hussain Fazle. On the identification of a vortex. Journal of Fluid Mechanics Digital Archive (JFM), 1995, 285: 69–94.
15. Lumley J L. The structure of inhomogeneous turbulent flows. Atmospheric Turbulence and Radio Wave Propagation, 1976, 166–178.
16. Sirovich L. Turbulence and the dynamic of coherent structures. I-Coherent structures. II-Symmetries and transformations. III-Dynamics and scaling. Quarterly of Applied Mathematics, 1987, 45: 561–571.
17. Chen H, Reuss D L, Hung D L S, Sick V. On the use and interpretation of proper orthogonal decomposition of in-cylinder engine flows. Measurement Science and Technology, 2012, 23: 085302.
18. Taira K, Brunton S L, Dawson S T M, et al. Modal analysis of fluid flows: An overview. AIAA Journal, 2017, 55(12): 4013–4041.
19. Schmidt O T, Colonius T. Guide to spectral proper orthogonal decomposition. AIAA Journal, 2020, 58(3): 1023–1033.
20. Schmid P J, Sesterhenn J. Dynamic mode decomposition of numerical and experimental data. Journal of Fluid Mechanics, 2010, 656(1): 5–28.
21. Premaratne P, Hu H. Analysis of turbine wake characteristics by using dynamic mode decomposition method. Denver: 35th AIAA Applied Aerodynamics Conference, 2017.
22. Mariappan S, Gardner A D, Kai R, et al. Analysis of dynamic stall using dynamic mode decomposition technique. AIAA Journal, 2013, 52(11): 2427–2439.
23. Kou J, Zhang W. Dynamic mode decomposition and its applications in fluid dynamics. Acta Aerodynamica Sinica, 2018, 36(2): 163–179.

24. Schmid P J. Dynamic mode decomposition and its variants. Annual Review of Fluid Mechanics, 2022, 54: 225–254.
25. Heinzel G, Rüdiger A, Schilling R. Spectrum and Spectral Density Estimation by the Discrete Fourier Transform (DFT), Including a Comprehensive List of Window Functions and Some New At-Top Windows. 2002.
26. Nekkanti A, Schmidt OT. Frequency-time analysis, low-rank reconstruction and denoising of turbulent flows using SPOD. Journal of Fluid Mechanics, 2021, 926: A26.

Chapter 5
High-Fidelity Simulation and Analysis of Flow in Compressor Blade Corner Region

5.1 Introduction

With the increase in compressor load, the three-dimensional separation phenomenon generated at the compressor blade root corner becomes more pronounced, gradually becoming one of the main factors affecting the aerodynamic performance such as efficiency and stability of the compressor. Suppressing the extent of corner separation through flow control technology is an important approach to improving compressor performance. The effectiveness of flow control strongly depends on the depth and accuracy of understanding the flow mechanisms. This understanding of complex physical mechanisms is largely built on the continuous development of research methods. However, due to the highly three-dimensional, unsteady, and complex interference mechanisms of the flow structures in the corner region, conventional Reynolds-Averaged Navier–Stokes (RANS) numerical methods and steady experimental techniques are insufficient for accurately capturing the intricate turbulent structures within the flow field. Additionally, conventional data analysis methods for flow variables struggle to identify the correlation mechanisms between turbulent structures in the complex coupled flow field. This greatly limits the understanding of the physical nature of corner separation, leading to a lack of unified knowledge about the detailed spatial flow field structures and their temporal evolution in the compressor corner region, as well as the causes of large-scale instabilities induced by corner separation.

Therefore, this chapter focuses on a high-load compressor cascade, employing DES and DDES numerical methods to conduct high-fidelity simulations of the corner flow field. Temporal and spatial decoupling techniques such as Dynamic Mode Decomposition (DMD) and Proper Orthogonal Decomposition (POD) are introduced into the analysis of the complex flow field of compressor corner separation. By examining the unsteady aspects, this study aims to gain a renewed understanding of the spatiotemporal evolution mechanisms of the corner flow field and to explore the causes of corner flow separation and instability.

R. Li et al., *Detached-Eddy Simulation Series Methods for Compressor Internal Flow*,
https://doi.org/10.1007/978-981-96-0255-1_5

5.2 Numerical Simulation of Corner Flow Field Based on DES-Series Methods

5.2.1 Controlled Diffusion Cascade

The research object is the NPU-A1 cascade designed by the Cascade Laboratory of Northwestern Polytechnical University using the controlled diffusion blade profile method. The geometry and parameters of the cascade are shown in Fig. 5.1 and Table 5.1. The cascade solidity is 2.30, and the geometric deflection angle is 39.6°, characterized by a narrow diffuser channel space and large curvature geometric surface. The research team has conducted extensive experimental research on this cascade, obtaining a wealth of experimental data, including angle of attack characteristics at different Mach numbers, isentropic Mach number distribution on the wall surface, and full-field pressure measurements on the channel wall surface.

The physical object is shown in Fig. 5.2. According to the existing angle of attack characteristics, the design condition of this cascade is an incoming flow Mach number of 0.5, an incoming flow angle of attack of 2.5°, and a Reynolds number of 7.16×10^5.

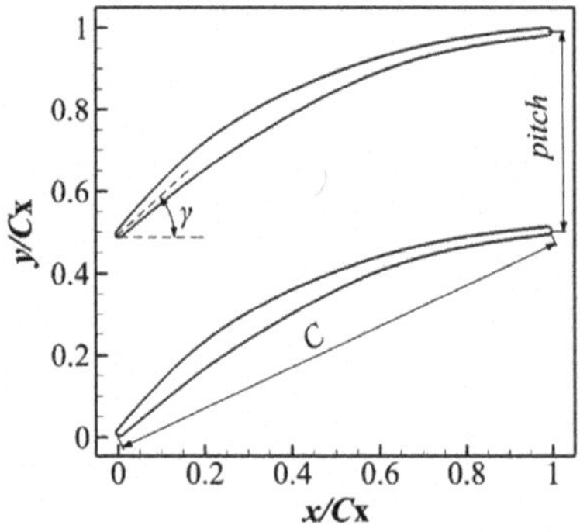

Fig. 5.1 Blade geometry and parameter definition

Table 5.1 Geometric parameters of NPU-A1 cascade

Parameter	Symbol	Value
Geometric inlet angle (°)	β_{1k}	45.8
Geometric outlet angle (°)	β_{2k}	6.2
Installation angle (°)	$\gamma_{cascade}$	26.6
Solidity (chord/pitch)	*C/pitch*	2.30
Aspect ratio (blade height/chord length)	*h/C*	1.43
Number of blades	*B_num*	8

Fig. 5.2 Picture of NPU-A1 cascade

5.2.2 Numerical Method

To reduce the computational load while considering symmetry, the computational domain is truncated from the endwall to the 50% blade height region, as shown in Fig. 5.3. The boundary condition at the 50% blade height is set as a mirror boundary. Since the corner separation of this cascade under the studied conditions does not extend to the mid-blade height, the flow at the mid-blade height still satisfies statistical laws, making this boundary feasible for DDES numerical simulations. A single passage is used for numerical simulation, with periodic boundary conditions on both sides of the passage. To mitigate potential disturbances from the inlet and outlet boundaries, the computational domain extends 1.5 and 2.8 times the axial chord length upstream and downstream of the leading and trailing edges, respectively. The grid is generated using AutoGrid5™, ensuring a y^+ below 0.5 except at the leading edge, where it is below 2.0.

The grid is prepared carefully. The computational domain is divided into RANS region (RR), focus region (FR) and departure region (DR). Requirements for grid number are different in those regions. In the FR, the grid number meets LES requirements which is $\Delta x^+ \leq 120$, $\Delta z^+ \leq 30$. In the DR and RR regions, the grid number is reduced. The region where large separation occurs is chosen as FR, contains from leading edge to one axial chord (C_x) length away from the trailing edge (shown in Fig. 5.4a), and in span-wise, it contains the region from end-wall to 40% blade height, as shown in Fig. 5.4b. The grid number in RR and DR regions are the same as those for RANS. The number of grid cell is given in Table 5.2. The total grid cell is 4.8 million.

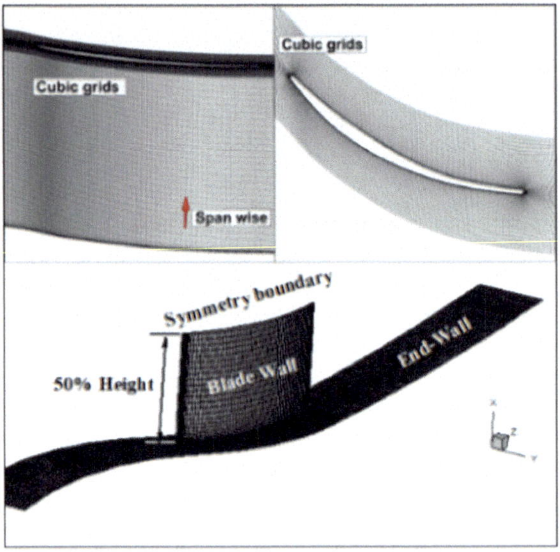

Fig. 5.3 Computational domain of NPU-A1 cascade

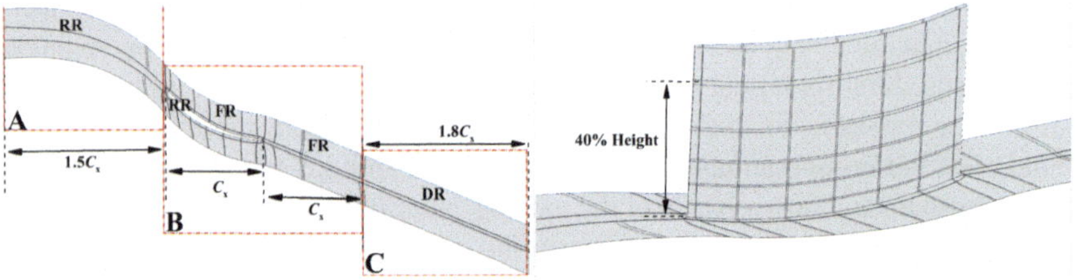

(a) B-B View (b) 3D View

Fig. 5.4 Schematic mesh delineation for numerical simulation of compressor blade grid based on DDES

Table 5.2 Grid division in different regions

Zone No.	A	B	C
Stream × span × chord-wise	20 × 154 × 94	296 × 154 × 94	20 × 154 × 94

5.3 Validation Based on Experimental Data

5.3.1 *Validation of DDES Method Accuracy for Simulating Compressor Blade Corner Region*

In this section, based on the experimental results, the accuracy of the DDES numerical results is verified qualitatively and quantitatively for the large-scale corner separation phenomenon. To minimize Riemann solver, Roe scheme is used for the convective flux calculation. The fifth-order weighted essentially non-oscillatory (WENO) scheme is used for reconstruction, and the minmod limiter is also applied for maintaining stability.

The fourth order dual time stepping is used for physical time marching. The selection of the physical time step should be related to the freestream velocity and the average grid scale ($\overline{\Delta}$). Their relationship is $\Delta t = \overline{\Delta}/V\text{max}$, where V_{max} is typically between 1.5 and several times the freestream velocity. Herein, the magnitude of the inlet velocity is approximately 174.5 m/s, and $Velocity_{\text{max}}$ is set as 2.2 times the inlet velocity. The average grid scale is approximated as $\overline{\Delta} = 5.1 \times 10^{-4}$ m for the whole grid region. Finally, $\Delta t = 1.34 \times 10^{-6}$ s is chosen in our simulation.

Numerical simulations are performed on a self-developed "cross-platform impeller machinery high-fidelity CFD software" [1]. In order to improve the efficiency of the non-stationary calculations, the program uses MPI-based parallel computing with 387 cores, and the calculations are performed on Tianhe-2.

The investigated operating condition is high subsonic with a large incidence, $Ma = 0.7$, $i = 5.0°$. Figure 5.5 shows a comparison between the experimental oil flow pattern on the suction surface and the limiting streamlines. From this comparison, it can be observed that:

(1) The initial location of separation on the suction surface: The experimental, RANS, and DDES results are 0.19*C*, 0.27*C* and 0.18*C* from the leading edge, respectively. Clearly, the DDES result is closer to the experimental data.
(2) The prediction of the recirculation region near the trailing edge: The RANS results show that the location of the singularity in the recirculation region

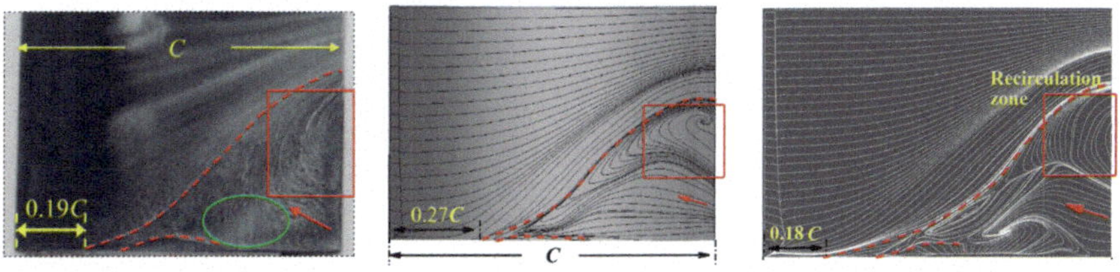

(a) Experimental results of oil flow (b) RANS (c) Time-averaged DDES results

Fig. 5.5 Experimental versus numerical comparison of the surface stream-lines on the suction surface of the blade ($Ma = 0.7$, $i = 5.0°$)

is within the blade passage near the trailing edge, whereas both the experimental and DDES mean results suggest that the singularity is outside the blade passage. Hence, the DDES results are more consistent with the experimental observations.

(3) The length of the vortex near the endwall corner region: As shown by the dashed line near the endwall in Fig. 5.5, the RANS result shows a relatively shorter vortex length, while the DDES result is closer to the experimental observation.

In the experiment, tracer particles in some regions of the suction surface were not blown away, as indicated by the green box in Fig. 5.5a. This is because the area is a low-velocity region, where the small velocity gradient results in lower wall shear stress, insufficient to overcome the adhesive force of the oil flow paint on the suction surface. However, if the viscosity of the paint is reduced, thus lowering its adhesion, the oil flow paint in other high-speed regions would be almost entirely blown away, and the results would be disturbed during the wind tunnel shutdown process. Considering this conflict, for future high subsonic compressor cascade experiments, it is advisable to use oil flow reagents of varying viscosities applied in different regions.

The quantitative comparison results are shown in Figs. 5.6 and 5.7, which illustrate the distribution of physical quantities inside and outside the passage, respectively. Within the passage (Fig. 5.6), the isentropic Mach number distribution at mid-span shows that the DDES method accurately captures the peak Mach number location and static pressure distribution. At the mid-span position at the passage exit, the Mach number distribution along the pitchwise direction is shown in Fig. 5.7, where the horizontal axis represents the dimensionless pitchwise coordinate. The measurement positions in the experiment are shown in Fig. 5.8, with the sensors located at 0.8 times the blade chord length from the passage exit. The results show that the DDES method has a good prediction capability for the peak position and the width of the wake, outperforming the RANS results, although there are slight differences near the suction side.

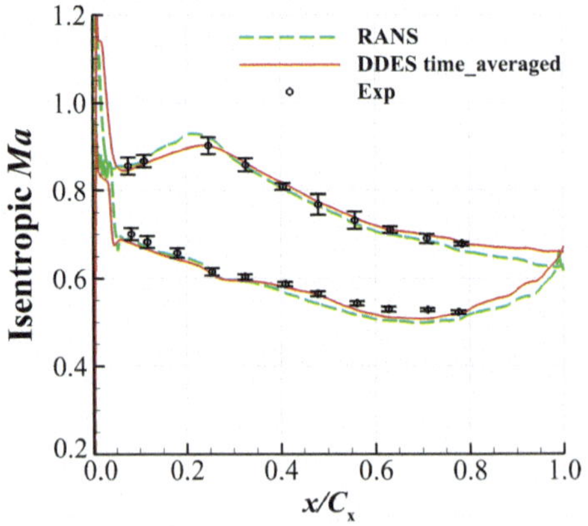

Fig. 5.6 Isentropic Mach number of blade midspan

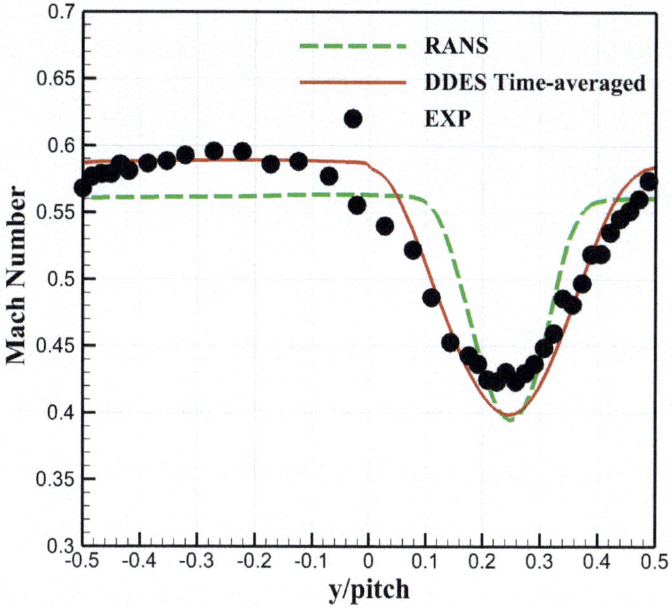

Fig. 5.7 Distribution of Mach numbers at intermediate blade heights at locations downstream from the trailing edge of the 0.8*C*

Fig. 5.8 Schematic of Mach number measurement locations

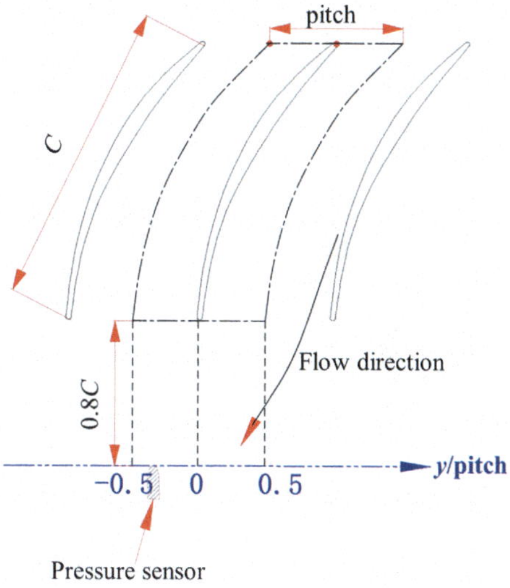

In summary, the DDES method accurately captures the extent of corner separation, the main secondary flows, and the primary parameters at mid-span, providing results that are closer to the experimental data compared to the RANS predictions.

5.3.2 Validation of DES for Simulating Compressor Blade Corner Region

This section will also use the NPU-A1 blade cascade as the research subject to perform numerical simulations for the condition of $Ma = 0.6, i = 5.0°$. By comparing the time-averaged results from DES with measurements from oil flow, pressure-sensitive paint (PSP), and pressure scan valves, the predictive capability of the DES method in the corner separation region is evaluated.

The previous section analyzed the advantages of DES-type methods in accurately simulating the flow field within compressor cascades, so no comparison with RANS is included here. The numerical simulations were conducted using the FINE/Turbo solver in the NUMECA software, with spatial discretization using a second-order central difference scheme and time advancement using a dual-time stepping method, with a physical time step of s and a maximum of 40 virtual time steps.

Figure 5.9 shows a comparison between the DES time-averaged limiting stream-lines on the suction surface and experimental oil flow results. The results indicate that the DES method captures the leading edge separation, corner separation, and the recirculation region near the trailing edge well compared to the experimental results. However, as with the previous cases, there is an area near the endwall where the low-speed region is not blown away.

Quantitative comparison results are shown in Figs. 5.10 and 5.11 which presents the isentropic Mach number distribution at different spanwise locations on the blade surface. The results from 22 static pressure taps show that the time-averaged DES

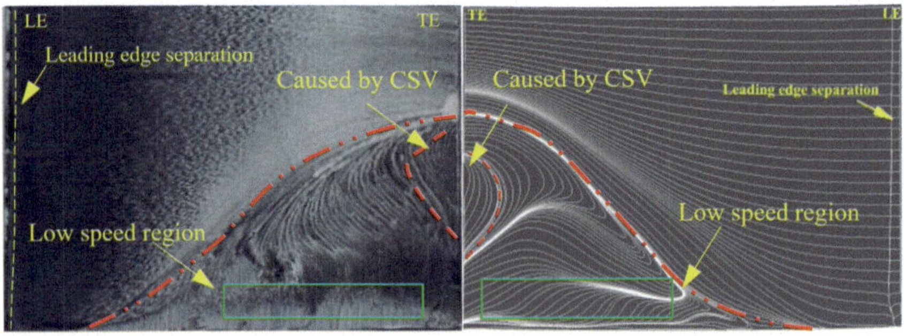

(a) Oil flow experimental results (b) DES time-averaged results

Fig. 5.9 Experimental versus numerical comparison of surface stream-lines at the suction surface of the blade ($Ma = 0.6$, $i = 5.0°$)

results match well with the experimental results, although there is a lack of validation near the leading and trailing edges due to measurement limitations. In recent years, pressure-sensitive paint (PSP) technology has gained widespread attention due to its high spatial resolution and minimal flow field interference. References [2–4] have applied PSP to measure the flow field in compressor cascades, providing valuable experimental data for high-resolution numerical validation. Figure 5.11 compares the PSP experimental results at 20% span with the time-averaged DES results, showing that the DES method accurately captures the isentropic Mach number distribution on the blade surface, consistent with the experimental data.

In summary, both qualitative and quantitative comparisons show that the DES-type method provides satisfactory results for the numerical simulation of the corner flow field in compressor cascades.

Fig. 5.10 Isentropic Mach number distribution at 50% blade heights

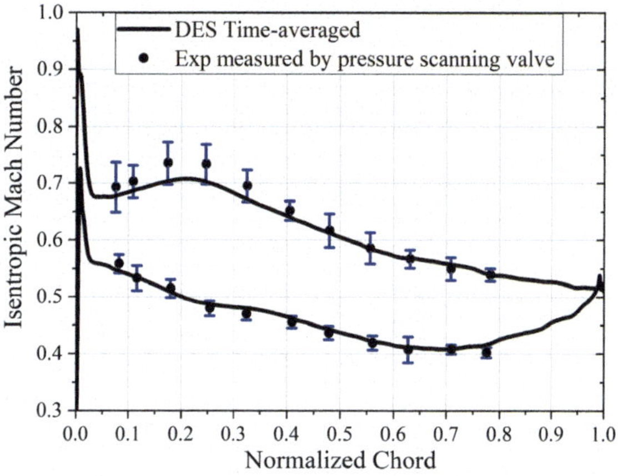

Fig. 5.11 Isentropic Mach number distribution at 20% blade heights

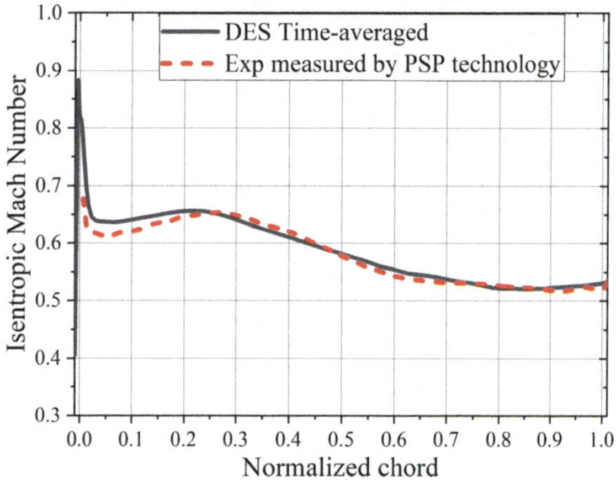

5.4 Complex Flow Field Analysis in the Corner

The greater impact on the efficiency and stability of a pressurized gas turbine is usually the turbulent structure where the pulsations are more non-constant and strong, i.e., the main pulsating turbulent structure. In this section, the DDES numerical method will be used to analyze the main pulsation region of the flow field outside the attached surface layer in the corner region of the pressurizer and to recognize the non-constant pulsation characteristics in this region. The numerical method and working conditions are consistent with those in Sect. 5.3.1.

5.4.1 *Turbulent Structure of Main Fluctuations In/Outside the Passage*

Research on the main fluctuating turbulent structures outside the boundary layer in the corner region first requires identifying the primary fluctuating areas. To quantitatively describe the magnitude of the fluctuations (unsteady effects), this section employs the parameter Su defined in Ref. [5], with the following definition:

$$Su(x, y, z) = \sqrt{\frac{1}{\Delta t} \int\limits_{t_0}^{t_0+\Delta t} \left[g(x, y, z, t) - \overline{g}(x, y, z)\right]^2 dt} \tag{5.1}$$

where $g(x, y, z, t)$ characterize the physical quantities of the flow field at a given transient, and $\overline{g}(x, y, z)$ are the corresponding time-averaged physical quantities of the flow field. Here, the pressure field is used for the calculations, and larger values of Su indicate stronger non-constant strength. The results are shown in Fig. 5.12, as can be seen in the figure:

(1) Within the blade passage, the unsteady fluctuating range is primarily concentrated between 10 and 25% of the blade height and is located near the suction surface.
(2) At the blade passage exit, the unsteady fluctuations are mainly concentrated in a ring-like region.

Next, we focus on analyzing the vortex structures corresponding to the main fluctuating regions. The Q-criterion is used to visualize the vortex structures in the flow field. The contour surfaces based on the Q-criterion at a certain instantaneous flow field, as shown in Fig. 5.13, provide an initial identification of the main flow structures. From the figure, we can observe:

- The largest-scale vortex structure in the blade passage is the passage vortex (PV), which forms near the leading edge and suction surface at the endwall. As the PV develops downstream, its influence range gradually increases.

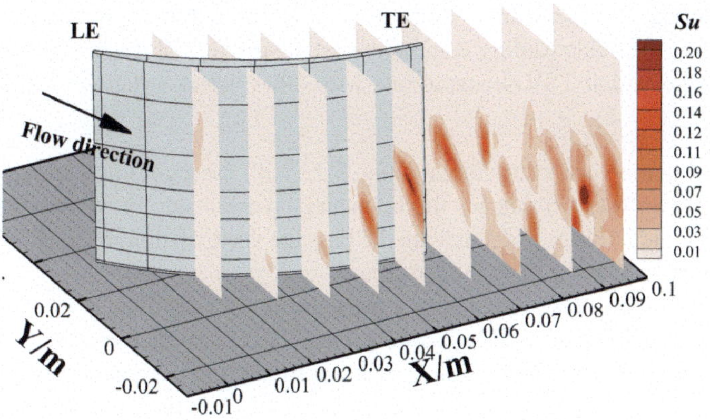

Fig. 5.12 Unsteadiness intensity at different planes [6]

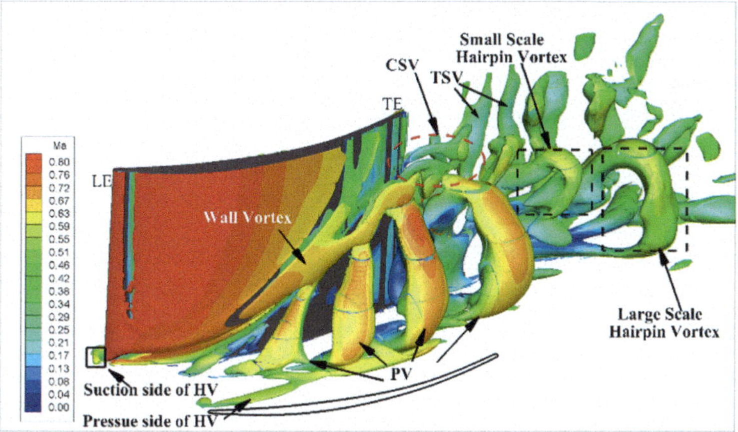

Fig. 5.13 Major vortex structures in transient flow field (Q-criterion isosurfaces and Mach number cloud, $Q = 7 \times 10^7$) [6]

- Due to the scraping effect of the passage vortex, wall corner vortices (WCV) or wall vortices (WV) form on the suction surface side of the blade.
- At the blade trailing edge, concentrated shedding vortices (CSV) are formed from the detachment of WCV.
- Horse vortices (HV) form near the leading edge close to the endwall. However, in this high-speed compressor cascade, the suction surface branch of HV is not very prominent.
- Additionally, downstream of the blade passage, large-scale and small-scale vortex structures alternate, as indicated by the black dashed box in the figure. These structures exhibit a hairpin shape and are referred to as hairpin vortices.

- The trailing-edge shedding vortex (TSV) is formed from the detachment of vortices at the blade trailing edge. The spatial structure of the instantaneous flow field suggests that TSV seems to mix with other vortex structures, such as CSV and PV.

Combining the vortex structures in Fig. 5.13 with the main fluctuating regions in Fig. 5.12, it can be seen that for the boundary layer outside the corner region, the primary unsteady vortex structures are the passage vortex within the passage and the hairpin vortex downstream of the passage.

An instantaneous flow field does not show the mixing and development process of vortex structures. Therefore, Fig. 5.14 presents vortex structure images at six typical moments obtained from the DDES simulation. Regarding the generation of the main vortex structures, PV originates from the endwall near the leading edge suction surface and is marked by the black solid line box in the figure. It gradually develops toward the middle of the blade as it progresses downstream. The six moments involve seven passage vortex structures, numbered from PV1 to PV7. As shown in Fig. 5.14a–f, the generation position of the passage vortex is quite stable and has been within the range of $x = 0.012\ \text{m} \pm 0.002\ \text{m}$ and $y = 0.016\ \text{m} \pm 0.004\ \text{m}$, consistently within the range of and, making it an ideal location for flow control.

In terms of the development of the main vortex structures:

- At time T1 (Fig. 5.14a), PV1 has developed to the blade passage exit. At time T2 (Fig. 5.14b), PV1 mixes with other vortex structures to form a smaller-scale hairpin vortex.
- At time T3 (Fig. 5.14c), PV2 reaches the passage exit and gradually mixes with other vortex structures. By time T5 (Fig. 5.14d), PV2 has developed into a larger-scale hairpin vortex.

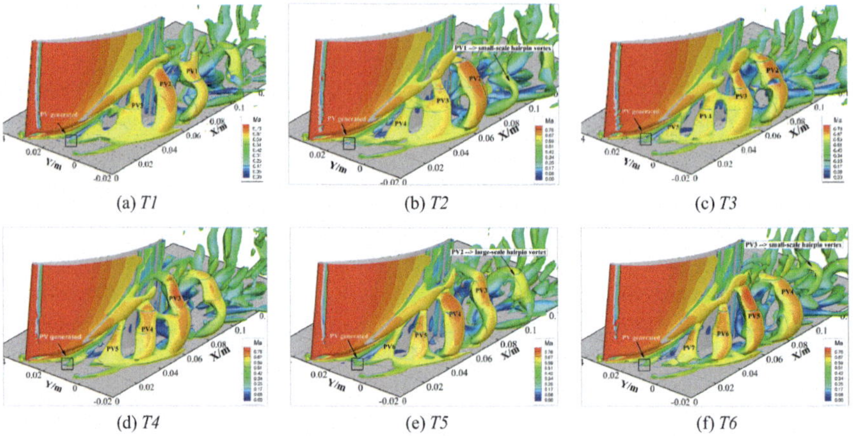

(a) *T1* (b) *T2* (c) *T3*

(d) *T4* (e) *T5* (f) *T6*

Fig. 5.14 Vortex development, iso-surface of 7×10^7 contoured with the Mach number. Taking $T1$ as the initial moment, the relationship between other moments and $T1$ is: $T2 = T1 + 56\Delta t$; $T3 = T1 + 98\Delta t$; $T4 = T1 + 158\Delta t$; $T5 = T1 + 218\Delta t$; $T6 = T1 + 356\Delta t$

- At time T6 (Fig. 5.14f), PV3, similar to PV1, forms a smaller-scale hairpin vortex, and this process continues in a cyclic manner.

In summary, the primary fluctuating structures in the compressor blade corner region are the passage vortex within the passage and the hairpin vortex at the passage exit. The formation position of the passage vortex is relatively stable, and at the passage exit, it mixes with other vortex structures to form hairpin vortices of alternating large and small scales.

5.4.2 Temporal and Spatial Evolution of Main Fluctuating Turbulent Structures Inside the Passage

Based on the analyses in the previous sections, a preliminary understanding of the main fluctuating regions and turbulent structures in the corner flow field has been established. However, due to the high coupling of turbulent structures, the formation mechanisms, temporal development, spatial structural changes, and mixing mechanisms of these turbulent structures are still not well understood. To address this, this section employs the Dynamic Mode Decomposition (DMD) method to temporally and spatially decompose the main fluctuating regions inside and outside the passage from the DDES numerical results, aiming to analyze the formation and evolution mechanisms of the main vortex structures in the corner region flow field.

To consider multiple physical quantities comprehensively, the DMD analysis includes velocity vectors and pressure. These quantities are arranged sequentially according to grid points to form an input matrix. In this study, the number of snapshots (N) is 520.

From the transient flow field vortex structure diagrams (Fig. 5.13), it can be observed that the main vortex structures inside the passage originate from the region close to the suction surface. Therefore, the analysis of vortex structures inside the passage should also focus on this region. Considering the large number of grid points and sample size, the memory requirements for performing a full-field DMD analysis of the corner region are substantial and challenging. Thus, a layered analysis approach is used, where cross-sectional slices are extracted at different blade heights from the selected region, and DMD decomposition is performed for each slice, as shown in the region of interest (ROI) in Fig. 5.15. Four cross-sections are extracted, corresponding to blade heights of 10%, 15%, 20%, and 25%, with spanwise coordinates of $Z = 0.02$ m, $Z = 0.025$ m, $Z = 0.03$ m, and $Z = 0.035$ m, respectively.

The relationship between the 2-norms of different modes and their frequencies after DMD decomposition is shown in Fig. 5.16. For a stable flow field, a larger 2-norm of a mode indicates a greater proportion of that mode in the unsteady flow field composition. The mode with a frequency of 0 represents the time-averaged flow field over the analyzed time range, labeled as Mode 0, which will not be analyzed in detail here. Typically, Mode 0 has the largest 2-norm among all modes. However, at the 10% blade height cross-section (Fig. 5.16a), due to proximity to the endwall

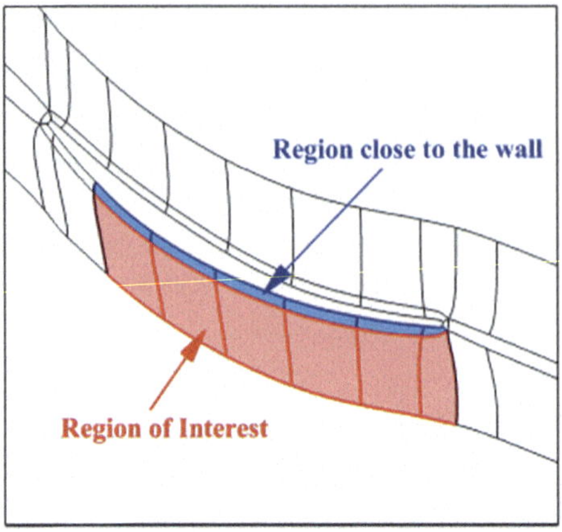

Fig. 5.15 DMD study area in the channel

and higher unsteady pulsation intensity, the 2-norm value of Mode 1 exceeds that of Mode 0. In the pulsating flow field, Mode 1 is the primary source of fluctuations. Below the 25% blade height (Fig. 5.16a–c), the 2-norm of Mode 1 is significantly greater than that of other unsteady modes. At the 25% blade height (Fig. 5.16d), the 2-norm values of multiple modes are nearly the same, indicating strong vortex mixing and energy transfer at this cross-section.

The reconstruction of the main fluctuating mode (Mode 1) at different cross-sections is shown in Fig. 5.17, which includes static pressure contours and velocity vector plots. Based on the location of the static pressure fluctuations and in conjunction with Fig. 5.13, this mode characterizes the dynamic behavior of the passage vortex as it progressively moves downstream. The direction of the pressure fluctuation trajectories indicates the development direction of the passage vortex, as shown by the arrows in the figure. It can be observed that the pressure fluctuation areas captured closer to the blade mid-span are nearer to the trailing edge, with trajectories increasingly biased toward the suction surface. It can be inferred that as the passage vortex moves downstream, its spanwise extent gradually increases. The closer movement trajectory towards the suction surface near the blade mid-span is caused by the increased circumferential pressure differential, as illustrated in Fig. 5.18.

Table 5.3 presents the growth rates and dimensionless frequencies (St) for Mode 1 at different cross-sections. All growth rates are very small negative values, indicating that Mode 1 is a stable, periodic dynamical mode. Additionally, *St* increases as the passage vortex develops and moves closer to the mid-span of the passage, suggesting that the movement speed of the passage vortex near the blade mid-span is greater than that near the endwall. This is one of the reasons why the frequency observed within the blade passage is typically a frequency band rather than a single frequency.

In summary, as the strongest unsteady vortex structure within the blade passage, the passage vortex gradually enlarges as it moves downstream. The movement trajectory closer to the blade mid-span is more influenced by the circumferential pressure differential, drawing it closer to the suction surface. Furthermore, due to the spatial

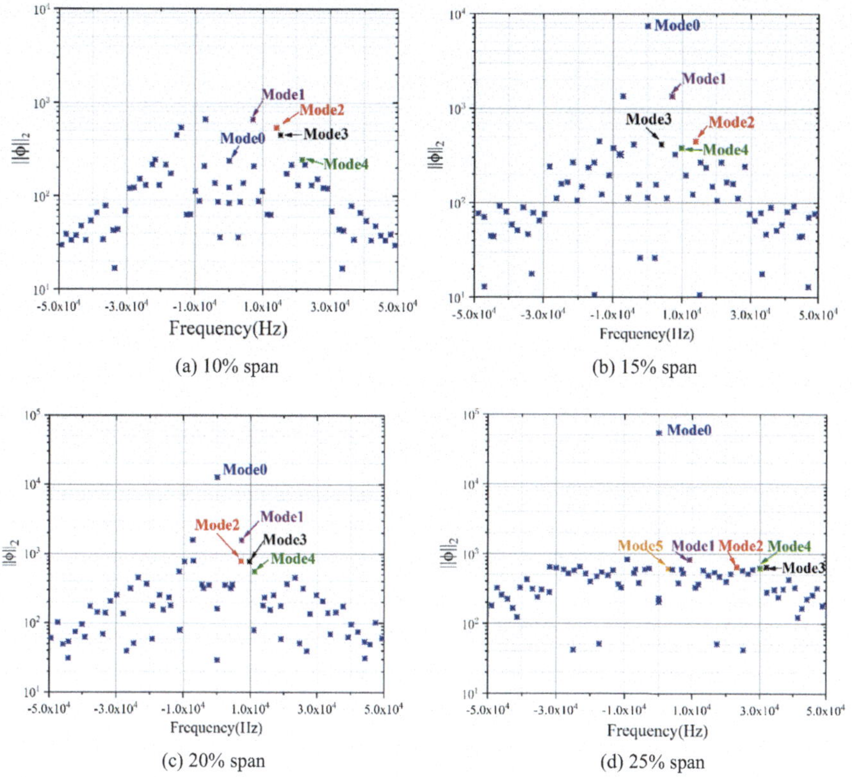

Fig. 5.16 Modal energy versus frequency at different blade span [6]

stretching effect of the vortex structures, there are slight variations in the frequency of the same vortex structure across different spatial regions.

5.4.3 Temporal and Spatial Evolution of Main Fluctuating Turbulent Structures Outside the Passage

Based on the previous section's analysis, it is evident that hairpin vortices are the predominant turbulent structures at the passage exit. This section will focus on the formation and development of hairpin vortices.

From Fig. 5.14, it is observed that hairpin vortices originate within the flow field range from $X = 0.07$ to 0.1 m. Thus, this section selects a cross-section perpendicular to the flow direction at $X = 0.085$ m (S3 plane) as the research area and performs DMD analysis. The results are shown in Fig. 5.19, where the first five modes are numbered from Mode 0 to Mode 4. Mode 0 represents the time-averaged flow field, with its reconstructed flow field results shown in Fig. 5.20, displaying the streamwise

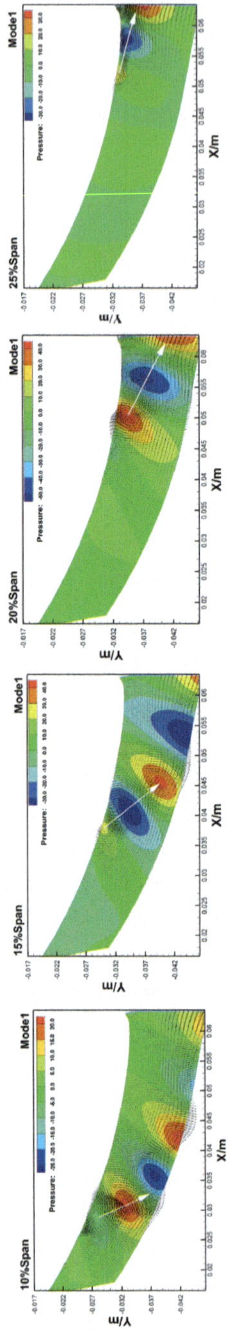

(a) 10% span (b) 15% span (c) 20% span (d) 25% span

Fig. 5.17 Spatial pattern of the dynamic mode at frequency of Mode 1 [6]

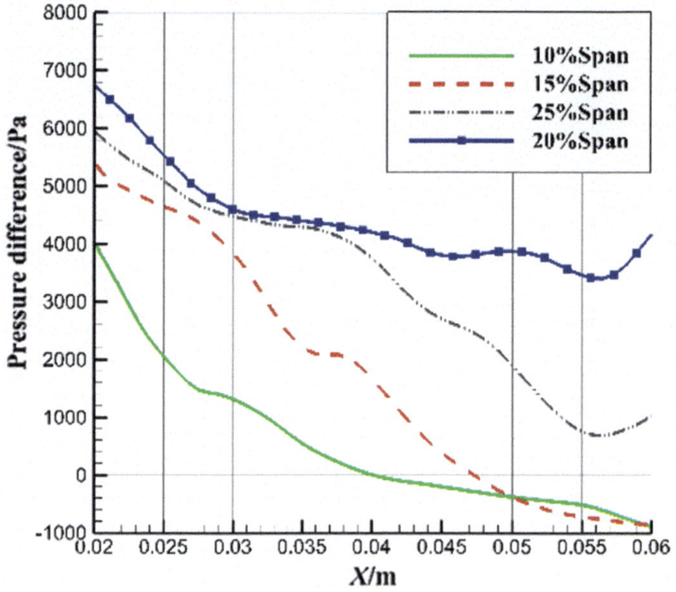

Fig. 5.18 Pressure difference between suction surface and middle of channel at different cross sections based on time-averaged results [6]

Table 5.3 Growth rates of Mode 1 at different blade spans

Position	10% span	15% span	20% span	25% span
St	1.213	1.230	1.259	1.602
Growth rates	-2.697×10^{-4}	-5.581×10^{-4}	-4.898×10^{-4}	-4.606×10^{-4}

velocity contour plot. The low-speed region in the figure represents the projection of vortices on this plane. Based on the spatial location of the low-speed region, it can be inferred that, in the range of $Z = 0.035$–0.06 m, the low-speed region is caused by the wake, while the large low-speed region at the exit is induced by the hairpin vortices.

To investigate the dynamic characteristics of this region, we further examine Modes 1 to 4, which represent dynamic features. From Fig. 5.19, it is evident that these four dynamic modes have significantly higher 2-norms compared to other modes, though the values are quite similar, indicating strong mixing and energy exchange in this area. Figure 5.21 shows the distribution of the real and imaginary parts of the DMD eigenvalues on the unit circle (dashed line), with the x-axis and y-axis representing the real and imaginary parts of the eigenvalues, respectively. Modes located on or inside the unit circle are stable modes, such as Mode 1, Mode 2, and Mode 3 in this case. These three modes are approximately on the unit circle, indicating that they represent periodic and stable flow fields. Modes outside the circle, such as Mode 4, are unstable.

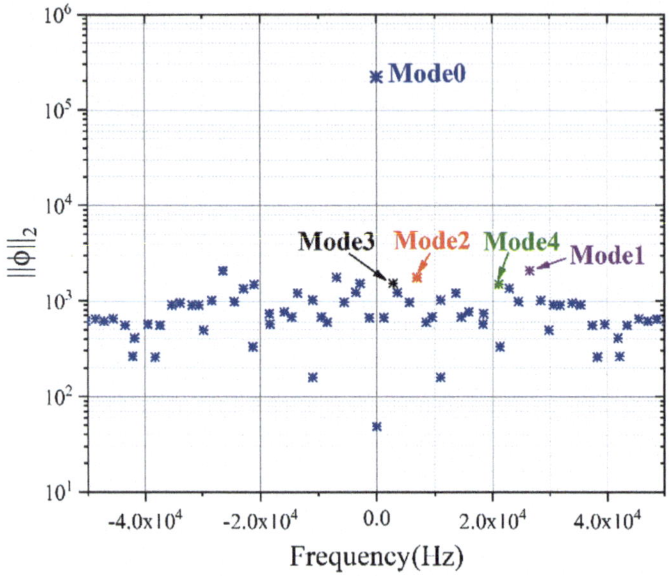

Fig. 5.19 DMD decomposition results at $X = 0.085$ m cross section. Mode 1: purple; Mode 2: red; Mode 3: black; Mode 4: green

Fig. 5.20 Mode 0 flow velocity cloud (view: from upstream)

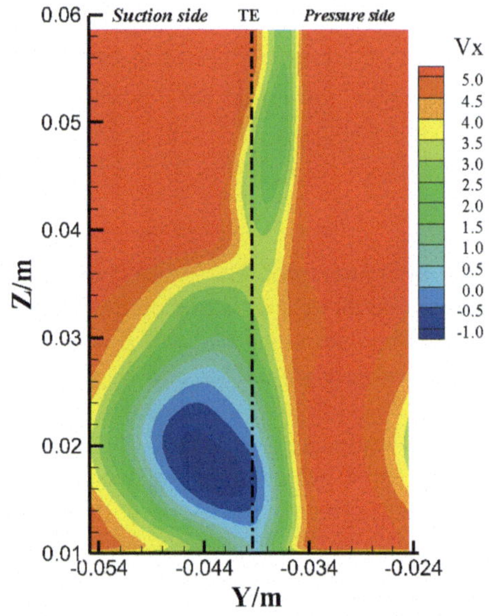

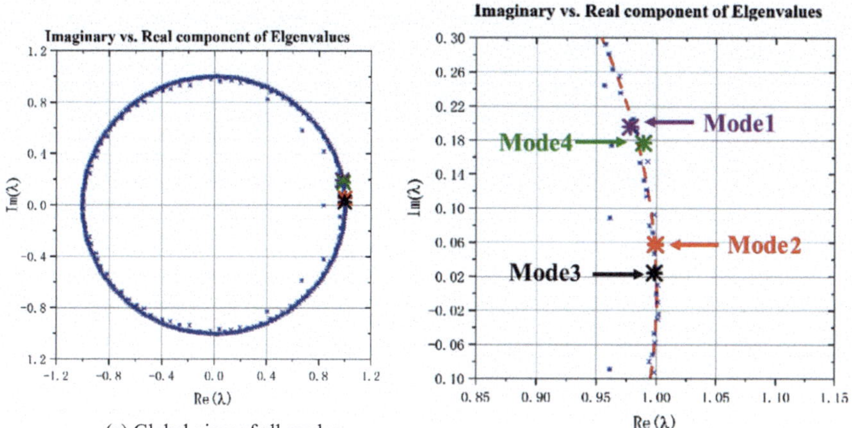

(a) Global view of all modes

(b) Layout enlargement

Fig. 5.21 Distribution of real and imaginary parts of modal eigenvalues at $X = 0.085$ m cross section on unit circle [6]. Mode 1: purple; Mode 2: red; Mode 3: black; Mode 4: green

The reconstructed flow fields for the stable modes Mode 1, Mode 2, and Mode 3 are shown in Fig. 5.22, with further calculations providing the vorticity contours. From the vorticity contours, it can be inferred that Mode 1 captures the frequency of the concentrated shedding vortex (CSV) passing through this section; Mode 2 captures the frequency of the interaction between the passage vortex (PV) and CSV; Mode 3 captures the frequency of the trailing-edge shedding vortex (TSV) passing through this section. Therefore, hairpin vortices are the result of the interaction between PV, CSV, and TSV. TSV develops into one leg of the hairpin vortex, while PV near the endwall develops into the other leg.

By comparing the frequencies of Mode 2 and Mode 3, as shown in Table 5.4, it can be observed that the frequency of Mode 2 is approximately twice that of Mode 3. This implies that the frequency of the passage vortex (PV) passing through

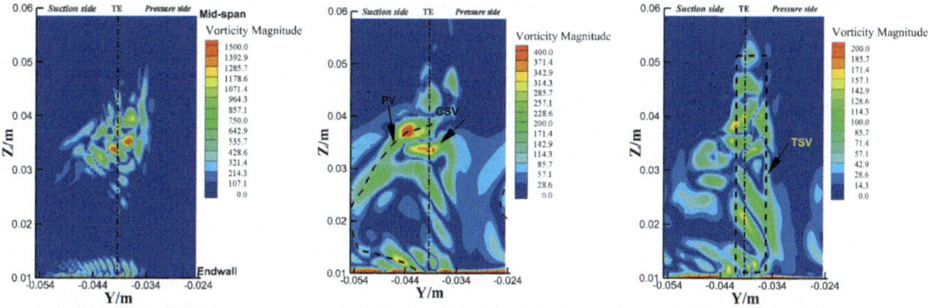

(a) Mode1

(b) Mode2

(c) Mode3

Fig. 5.22 Spatial pattern of dynamic mode at three modes in plane of $X = 0.0.85$ m (viewed from upwind direction) [6]

Table 5.4 Dimensionless frequencies of different modes after DMD decomposition at $X = 0.085$ m

Dynamic modes	Mode 1	Mode 2	Mode 3
St	3.26	0.85	0.44

this section is approximately twice that of the trailing-edge shedding vortex (TSV) passing through the same section. Therefore, when the passage vortex and the trailing-edge shedding vortex meet at the passage exit, they form large-scale hairpin vortices. Simultaneously, when adjacent passage vortices do not encounter the TSV near the trailing edge, they form smaller-scale hairpin vortices. This explains the alternating scales of hairpin vortices observed at the passage exit.

5.5 Instability Mechanism Analysis of Corner Separation

To investigate the underlying causes of flow instability in the three-dimensional corner region, the NPU-A1 blade row is chosen as the study subject. The DES numerical method is employed to simulate the flow at half blade height. Detailed numerical methods are discussed in Sect. 5.3.2 and will not be repeated here.

The angle-of-attack characteristics curve for the NPU-A1 blade row at a Mach number of 0.6 is shown in Fig. 5.23 (experimental results). The horizontal axis represents the incoming flow angle of attack (i), while the left vertical axis and the box in the figure indicate the static pressure ratio (π_c), and the right vertical axis and circles represent the total pressure loss (ϖ). It can be observed that the total pressure loss remains relatively constant until the angle of attack reaches 5.0°, after which it begins to increase sharply. For the pressure ratio, it continues to increase up to an angle of attack of 7.5°, after which it suddenly decreases, indicating a stall condition and transition to an unstable state. Therefore, an angle of attack of 7.5° is considered close to the stall condition. This section focuses on the case of $Ma = 0.6$ and $i = 7.5°$, with a stable condition of $Ma = 0.6$ and $i = 5.0°$ selected for reference, to explore the causes of instability through comparison.

5.5.1 *Analysis of Flow Field Structure Based on Time-Averaged Flow Field*

In the time-averaging of unsteady flow fields, the process typically uses the primary unsteady frequency (denoted as f_{main}) as a reference. This involves selecting two or more cycles of the flow field for time-averaging. In this study, f_{main} is defined as the frequency with the maximum energy in the FFT transformation of the unsteady fluctuations in the mass flow rate at the outlet. Time-averaging is performed using results from three converged cycles, as follows:

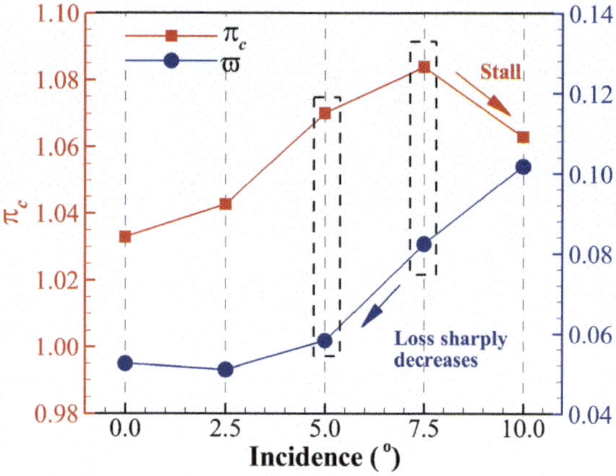

Fig. 5.23 Experiment results of incidence characteristics with Mach number of 0.6

$$\overline{g}(x, y, z) = \int_{t_{\text{start}}}^{t_{\text{start}}+3/f_{\text{main}}} g(x, y, z, t)dt \tag{5.2}$$

When the operating condition changes, the increase in flow losses and the occurrence of unstable pulsations typically result in alterations to the flow details. Therefore, this section will first investigate the flow topology.

The direction of the shear forces on the blade surface is influenced by the three-dimensional separation in the corner region. Figure 5.24 presents the limit streamline plots for the suction surface under stable conditions (incidence = 5.0°) and near-stall conditions (incidence = 7.5°), based on time-averaged results. Comparison reveals that, apart from differences in the size of secondary flow regions (including the corner separation and leading-edge separation zones), there is a vortex structure with a component perpendicular to the suction surface under near-stall conditions, denoted as "SSV (suction-surface separation vortex)" and highlighted in the dashed box in the figure.

In order to further analyze the difference in the details of the spatial flow field between the two cases, the analysis is carried out here with the help of the parameter of dimensionless helicity (normalized helicity, H_{n}) [8]. H_{n} is the velocity vector and vortex vector dot product, viz.:

$$H_{\text{n}} = \frac{\mathbf{W} \cdot \boldsymbol{\psi}}{|\mathbf{W}||\boldsymbol{\psi}|} \tag{5.3}$$

From the definition, H_{n} is a scalar that characterizes the degree to which the velocity vector is parallel to the vortex vector and can assist in determining the vortex core position and the direction of the vortex vector. The definition assumes that when the velocity vector is parallel to the vortex vector, it represents the position of the vortex core. Thus, when the value of H_{n} is closer to − 1 or 1, it indicates proximity to the vortex core. The sign of H_{n} indicates whether the direction of the vortex vector

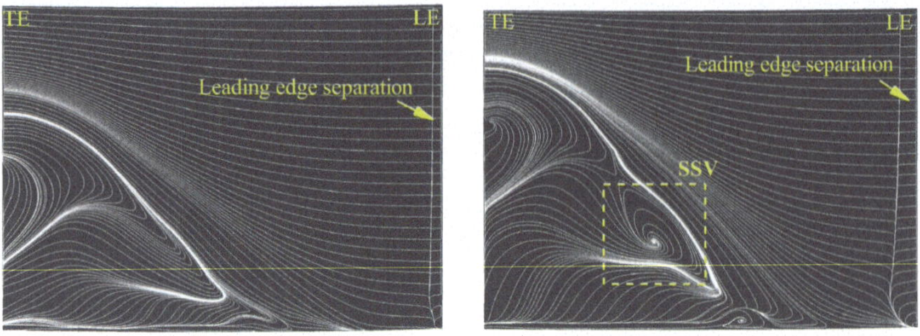

(a) incidence = 5.0° (stable) (b) incidence = 7.5° (near-stall)

Fig. 5.24 Limit streamlines on suction surface with different incidences from time-averaged DES results [7]

is aligned with or opposite to the direction of the velocity vector. Therefore, by considering the magnitude and sign of H_n along with the spatial location, the type of vortex can be identified.

The non-dimensional helicity cloud plot within the channel is shown in Fig. 5.25, along with the vortex cores of the main vortex structures in the flow field, indicated by the green lines. Different vortex structures have different signs for H_n. For instance, for the corner vortex (CV), the vortex vector near the endwall points towards the upstream recirculation zone, so both the velocity vector and vortex vector point upstream, resulting in a positive H_n. For the concentrated separation vortex (CSV) recirculation zone, the vortex vector has a downstream component while the velocity vector points upstream, leading to a negative H_n. For the passage vortex (PV), both the velocity vector and vortex vector point downstream, so H_n is positive. Other vortex types, such as the wall corner vortex (WCV) and horseshoe vortex (HV), follow similar logic, as summarized in Table 5.5.

Combining the vortex cores and H_n distribution, it can be observed that under the stable condition (incidence = 5.0°) (Fig. 5.25a), there are distinct vortex cores for leading-edge separation vortex, concentrated shedding vortex, corner vortex, and the pressure side branch of the horseshoe vortex in the blade passage. At the Plane 4 section, the pressure side branch of the horseshoe vortex intersects with the corner vortex, suggesting potential interference as they continue downstream. Under the near-stall condition (incidence = 7.5°) (Fig. 5.25b), in addition to the vortex cores for leading-edge separation vortex, corner vortex, concentrated shedding vortex, and pressure side branch of the horseshoe vortex, there is also a suction surface separation vortex (SSV) starting from the blade suction surface and disappearing between Plane 4 and Plane 5. In the upstream flow field of the SSV, the flow field structures are similar for both conditions, but the appearance of the SSV causes the vortex core of the pressure side branch of the horseshoe vortex to disappear after Plane 4 (Fig. 5.25b, top right inset), indicating that the presence of SSV intensifies the mixing effects in the flow field. Based on the analysis of the time-averaged flow field details, the

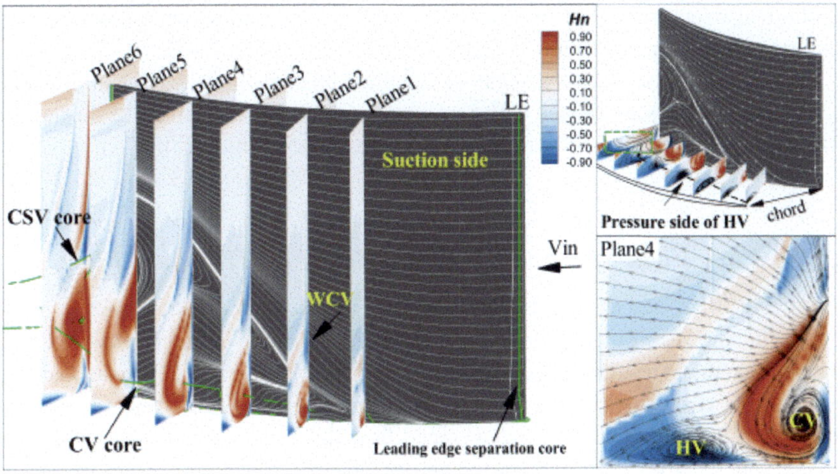

(a) incidence = 5.0° (stable)

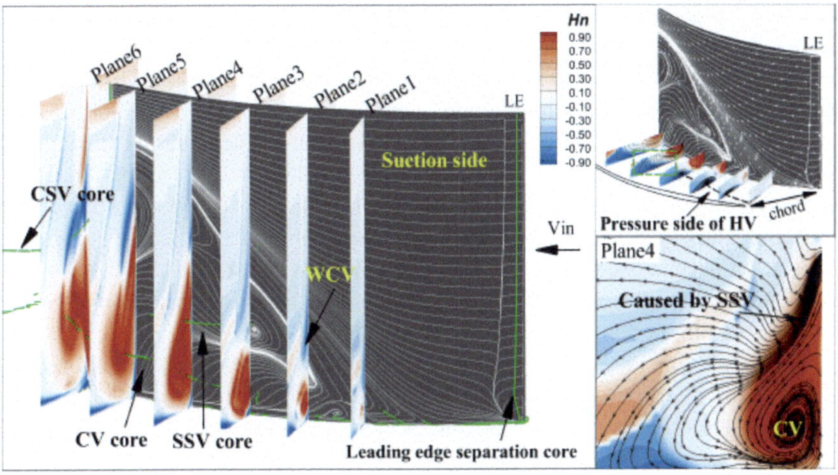

(b) incidence = 7.5° (near-stall)

Fig. 5.25 Contours of H_n and vortex core in passage, based on time-averaged DES results [7]

Table 5.5 The sign of H_n of different types of vortexes in the corner separation region

Vortexes type	CV	CSV	WCV	PV	HV suction branch	HV pressure branch
Sign of H_n	+	−	+	+	+	−

topological structures of the high-subsonic blade passage flow field under the two conditions are summarized in Fig. 5.26.

Based on the above analysis, it can be concluded that the flow field instability and increase in flow loss are related to the presence of the SSV. Therefore, the formation and development of the SSV will be further investigated (Fig. 5.27).

5.5.2 *Flow Field Structure Analysis Based on Transient Flow Field*

The time-averaged results smooth out the transient pulsations in the flow field, making it impossible to represent the unsteady dynamic characteristics caused by corner separation. Furthermore, a single time-averaged flow field may correspond to multiple unsteady flow fields. This section will analyze the unsteady characteristics of corner separation from the perspective of transient flow fields, with a particular focus on the formation and development of suction-surface separation vortex (SSV), which shows significant differences between stable and unstable conditions.

According to Délery's topological theory [9], a point is considered a separation point if streamlines enter it, while a point is considered an attachment point if streamlines exit it. Figures 5.28 and 5.29 illustrate the suction-surface limiting streamlines for incidence angles of 5° and 7.5°, respectively. It should be noted that, theoretically, streamlines and streamlines coincide in a time-averaged flow field. However, in a numerically discretized flow field, a numerical solution at a specific moment is effectively the time-averaged result within a discrete time step. Thus, the limiting streamlines at a specific moment can still qualitatively show the projection of vortex structures on the suction surface over short periods.

It is evident that the SSV is present in both flow fields. For both conditions, the corner vortex's vortex vector points upstream, indicating that a strong recirculation driven by the corner vortex develops from the blade trailing edge upstream, as shown in the region near the end and trailing edge in Fig. 5.28. This recirculation first attaches to the blade suction surface, then is compressed downstream by the wall corner vortex (WCV), and subsequently separates again from the blade suction surface, forming the suction-surface separation vortex (SSV).

In the stable condition, the SSV develops downstream driven by the wall corner vortex and mixes with the concentrated separation vortex (CSV) (Fig. 5.28c), eventually being expelled from the channel with the development of CSV (Fig. 5.28d). Since the SSV periodically forms, moves downstream, and is expelled from the channel, it is not captured in the time-averaged flow field. However, in the near-stall condition, the SSV persists within the channel after formation (Fig. 5.29a–d), with only slight changes in the vortex core position over time. After separating from the suction surface, the SSV does not mix with the CSV but instead interacts with the corner vortex.

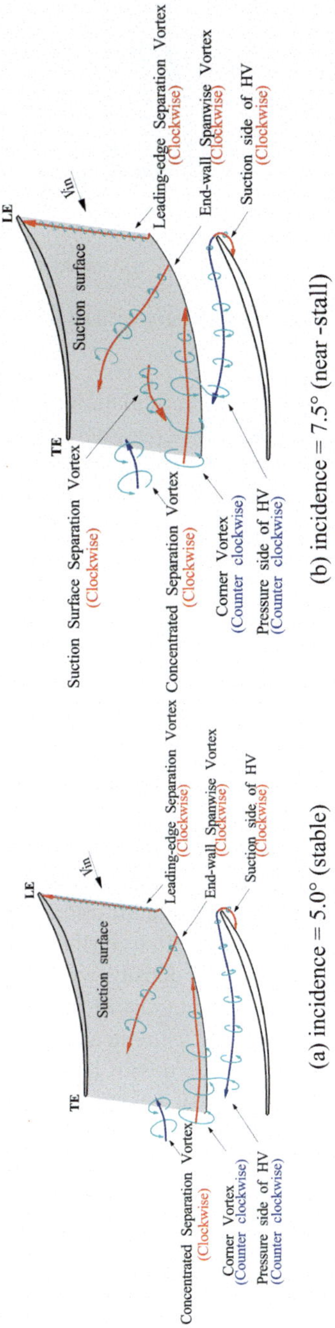

Fig. 5.26 Topological structure under different conditions [7]

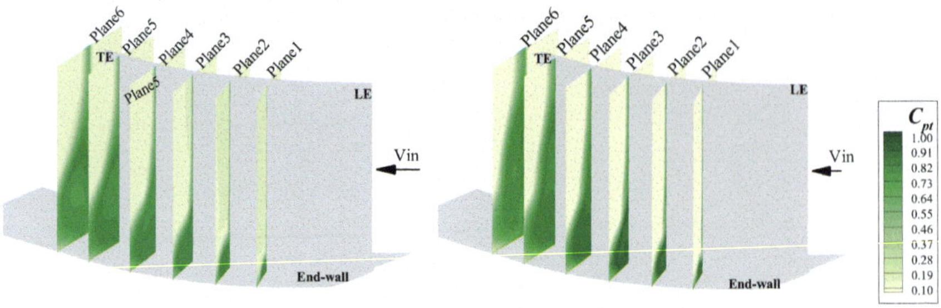

(a) incidence=5.0°(stable) (b) incidence=7.5°(near-stall)

Fig. 5.27 Contours of C_{pt} in passage based on time-averaged DES results [7]

In summary, the SSV is formed by the recirculation driven by the corner vortex attaching to the suction surface and then re-separating. The SSV is present in both stable and near-stall conditions, but differs in that, under stable conditions, it mixes with the CSV and is expelled from the blade channel, whereas, in near-stall conditions, it mixes with the CV and remains within the channel. The next step is to further explore how this unsteady characteristic of the SSV affects the flow field stability using the DMD analysis method.

5.5.3 *Analysis of Flow Instability Mechanisms*

Considering the large number of grid nodes and temporal snapshots, applying DMD analysis to the entire domain would require too much memory, making it impractical. Therefore, this section selects a specific area as the research domain. To analyze the corner separation phenomenon more clearly from a three-dimensional perspective, the research domain is chosen to be a 3D region, differing from previous work using DMD analysis [10, 11]. As shown in Fig. 5.30, to further reduce the computational load of the DMD decomposition, the grid density within the research domain is reduced by one-third in the I, J, and K directions of the computational coordinate system when extracting data. The sampling frequency is 125,000 Hz. The DMD analysis program is executed on a workstation with 192 GB of memory.

DMD decomposition results in a series of modes, each corresponding to a fixed oscillation frequency and growth rate. The 2-norm of a DMD mode is typically used to determine the modes that need to be studied. It is generally believed that modes with larger 2-norms are more significant to the unsteady nature of the flow field. The relationship between the dimensionless frequency *St* and the 2-norm is shown in Fig. 5.31. The larger the 2-norm, the more significant the mode's effect on the initial evolution of the unsteady flow field. Another parameter to consider is the growth rate. If the growth rate is negative, it indicates that the amplitude of the unsteady oscillation will gradually decrease over time, diminishing its effect on the later stages

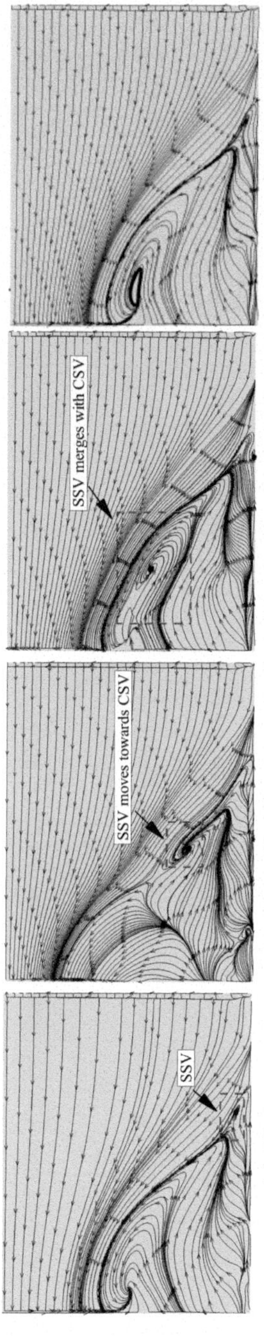

Fig. 5.28 Transient suction surface limit flow lines for four moments under stabilized conditions (incidence = 5.0°) [7]

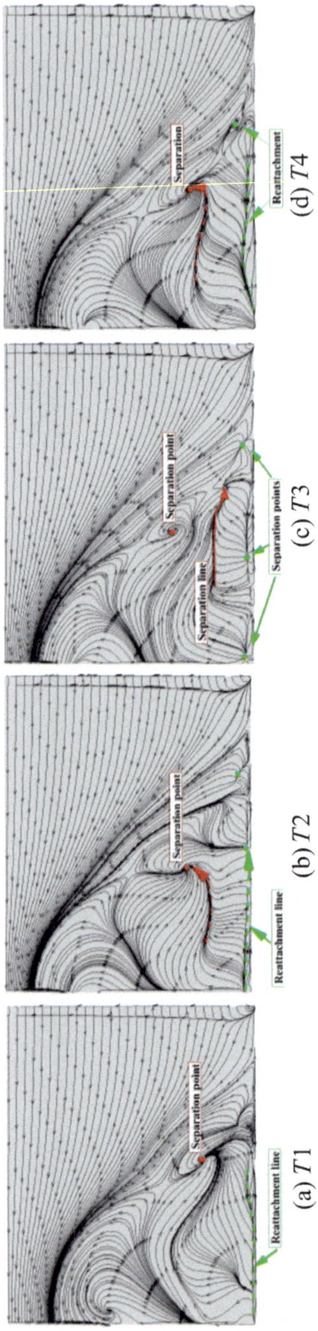

Fig. 5.29 Transient suction surface limiting flow lines for four moments in near-stall condition (incidence = 7.5°) [7]

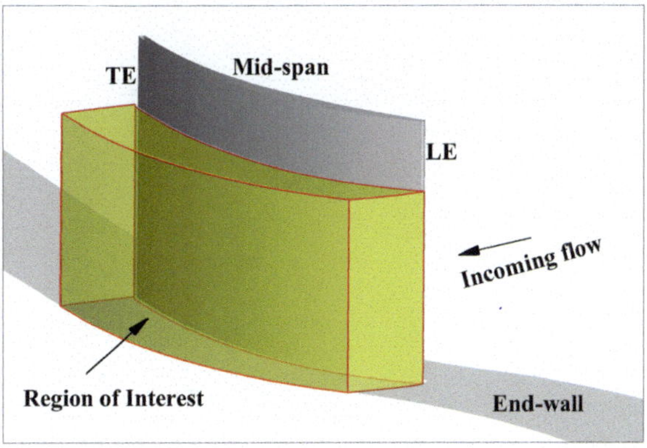

Fig. 5.30 Region of interest (3D volume) in DMD calculation

of evolution. If the growth rate is close to zero, it suggests that the mode exists stably and periodically in the flow field. Conversely, a positive growth rate indicates that the oscillation amplitude will gradually increase, playing a decisive role in the flow field ultimately becoming unstable. Mode 0 represents the time-averaged result without oscillations and is not discussed here.

In stable conditions (Fig. 5.31a), modes with larger 2-norms are concentrated at low frequencies. However, in near-stall conditions (Fig. 5.31b), there are two peaks, indicating that modes with larger 2-norms appear not only at low frequencies but also at high frequencies. This section first analyzes the stable modes: Table 5.6 shows the growth rates of several typical modes to analyze their impact on the flow field. Their growth rates are close to zero, meaning that Mode 1 to Mode 4 are stable modes, which conversely indicates that this condition is stable.

For the near-stall condition, we first analyze the growth rates of high-frequency modes. As shown in Fig. 5.32, the relationship between the growth rates and frequencies for all modes with a 2-norm greater than 500 is depicted, where the size of the spheres represents the magnitude of the 2-norm. Although the high-frequency modes have larger 2-norms, all the mode growth rates are less than zero. This indicates that, over time, the oscillations of high-frequency modes will gradually decay or remain unchanged in their periodic oscillations, thus not having a decisive impact on the stability of the flow field. Upon reconstructing the flow fields of all the mentioned high-frequency modes, it was found that high-frequency oscillations occur near the endwall at the trailing edge of the blade. This section uses Mode 1 as an example to illustrate this phenomenon, with the reconstructed pressure field shown in Fig. 5.33. Since this phenomenon only appears in the near-stall condition, this characteristic can be used for early warning of flow field instability.

However, for the low-frequency modes in the near-stall condition, three modes with relatively large 2-norms, marked as Mode A, Mode B, and Mode C in Fig. 5.31c, have growth rates listed in Table 5.7, showing significantly positive values. This indicates that these three modes play an important role in the stability of the flow field. Therefore, it is necessary to reconstruct the flow fields of these three modes and conduct an in-depth analysis of the corresponding physical phenomena.

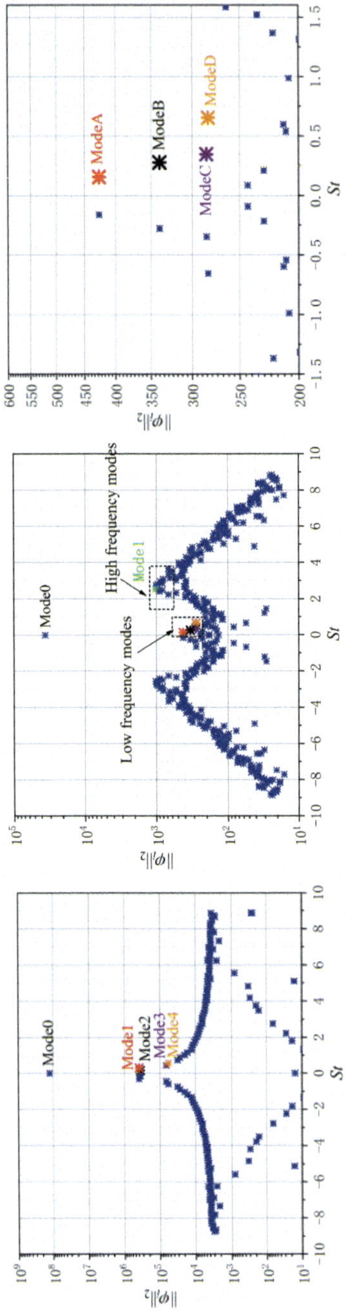

(a) stable condition (incidence = 5.0°)

(b) near-stall condition (incidence=7.5°) all frequency modes

(c) near-stall condition (incidence=7.5°) low frequency modes

Fig. 5.31 DMD spectra under difference incidences [7]

Table 5.6 Growth frequencies of different modes for stable conditions (mode numbers as in Fig. 5.31a)

Mode number	*St*	Growth rate	Stable condition
1	0.65	-4.43×10^{-8}	Stable
2	0.15	1.00×10^{-7}	Stable
3	0.44	1.60×10^{-7}	Stable
4	0.58	-1.72×10^{-8}	Stable

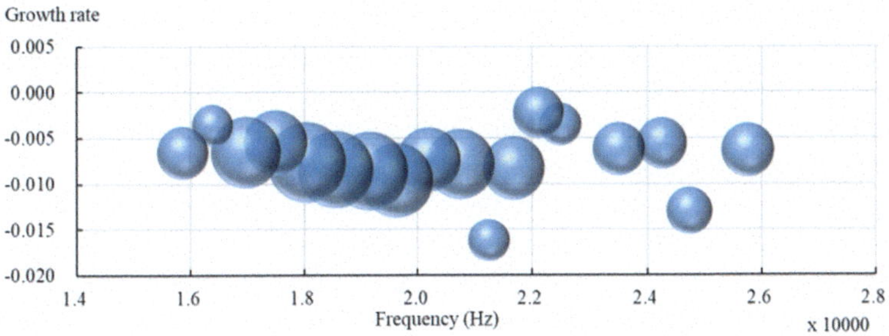

Fig. 5.32 Growth rate of a high-frequency mode for near-stall conditions, with the sphere size representing the 2-parameter size of the mode [7]

Fig. 5.33 Contours of Mode 1 at near-stall condition, where the non-green colors represent areas of high fluctuation [7]

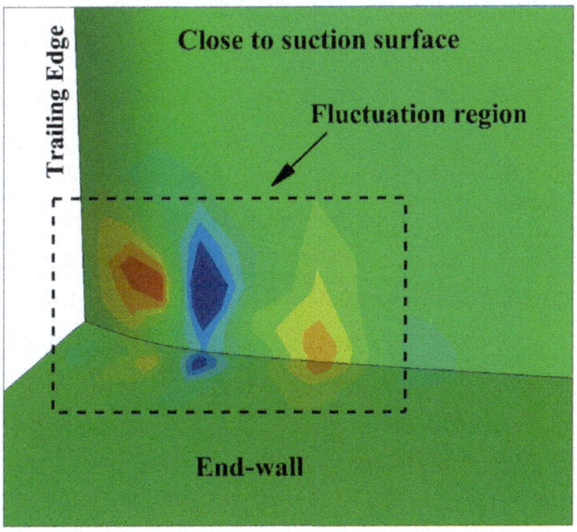

Table 5.7 Frequency and growth rate of different modes for near-stall condition, see Fig. 5.31c for mode numbers

Mode number	Frequency (Hz)	Growth rate	Stability
A	1143	4.95×10^{-3}	Unstable
B	2000	2.95×10^{-3}	Unstable
C	2500	1.97×10^{-3}	Unstable
D	4642	-1.05×10^{-3}	Stable

The primary fluctuation regions of the three modes are illustrated in Fig. 5.34 using pressure isosurfaces. The isosurface values are set at 50% of the maximum pressure fluctuation amplitude for each mode. The contour plots represent the dimensionless streamwise velocity Normalized_v_z, with the contour lines indicating the static pressure near the suction surface. Each mode has an equal number of contour lines.

For Mode A (Fig. 5.34a), the high fluctuation region occurs in the corner area formed by the suction surface and the endwall, primarily concentrated below 25% of the blade height. This coincides with the locations where the suction-surface separation vortex (SSV) and corner vortex (CV) form, suggesting that this mode represents the interaction between the CV and SSV. Similarly, for Mode B (Fig. 5.34b), the main fluctuations occur near the middle of the blade passage and close to the endwall, indicating that this mode represents the interaction between the CV and the pressure-side branch of the horseshoe vortex (HV). For Mode C (Fig. 5.34c), the main fluctuation region includes two parts: one near the corner area and the other near the location where the concentrated separation vortex (CSV) forms, suggesting that this mode represents the interference between the CV and CSV.

The interactions among these vortex structures create the primary sources of unsteady fluctuations in the flow field. Notably, the mode representing the interaction between the CV and SSV (Mode A) has the largest 2-norm and growth rate, playing a decisive role in the instability of the flow field. Therefore, we will focus on analyzing Mode A in detail next.

According to the time-averaged flow field shown in Fig. 5.24, the suction-surface separation vortex (SSV) originates at approximately 20% of the spanwise height, corresponding to a spanwise position of 96.0 mm. Therefore, in this section, the flow field for Mode A is reconstructed, and a cross-sectional analysis at this position is performed. The results, depicted in Fig. 5.35, illustrate the pressure field and velocity vector field at four representative moments within one cycle.

At phase = 0 in Fig. 5.35, the surface separation flow separates from the suction surface of the blade near $z = 37.5$ mm (streamwise), which is the position the SSV occurs. After separation, the SSV meets with a backflow caused by the CV. At phase =T/2, the backflow is cut into two parts by the SSV. One part moves forward towards to the leading edge, and the other part mixes with the SSV and finally causes the disappearance of SSV close to trailing edge; at phase π, the SSV is blocked by the enhanced backflow caused by the CV, so the SSV is pushed upstream and towards the mid-passage, which can lead to a wider range of blockages. At phase = 3T/4, driven by the main stream, the blockage of the passage is weakened.

Therefore, we can conclude that the mixing of the corner vortex and other vortices is the main cause of instability in the flow field, and among the vortices, the periodic mixing of the CV and SSV is the key factor for instability. The backflow and the downstream fluid, generated by the corner vortex and the suction surface separation vortex, respectively, cause a blocking effect. Two vortex structures develop near the middle of the passage and enlarge the low-energy area, which leads to the blade stall. Therefore, to widen the stability of the blade, the flow field of the suction side of the blade should be controlled to suppress the generation of an SSV or push the SSV out of the passage.

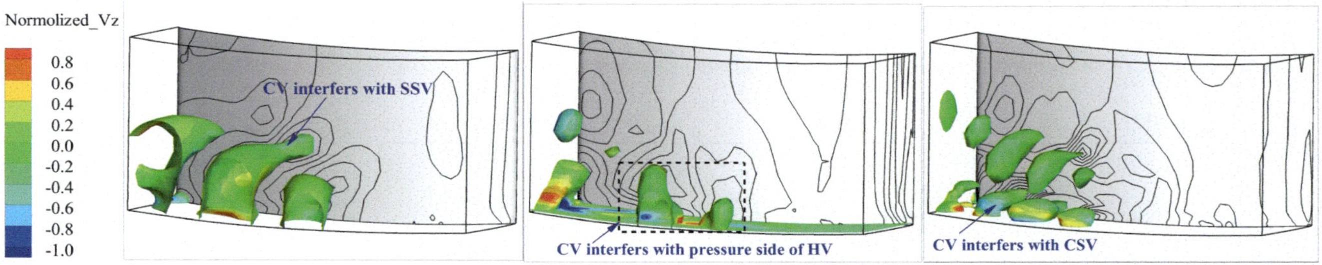

(a) Mode A (b) Mode B (c) Mode C

Fig. 5.34 Reconfiguration of the flow field for low-frequency modes for near-stall conditions [7]. Equivalent surfaces: pressure; clouds: dimensionless flow velocity; contours: static pressure values near the suction surface

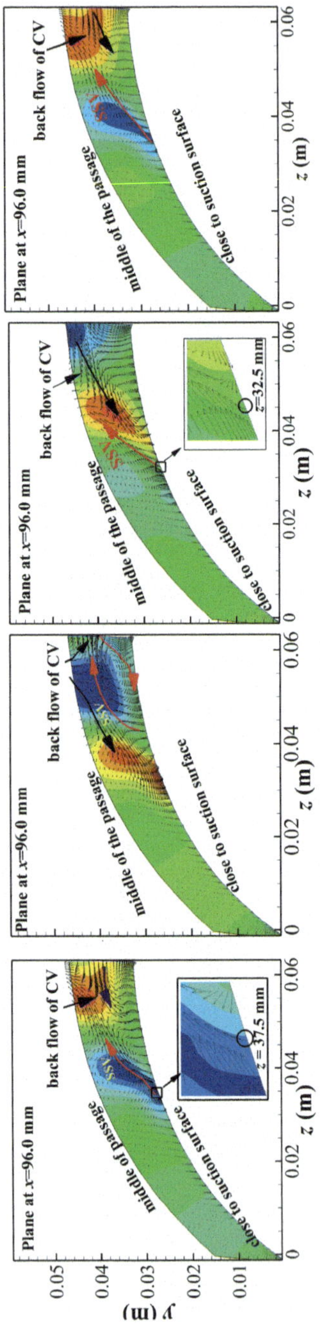

Fig. 5.35 Spatial pattern of the dynamic mode at frequency of Mode A at near stall condition (counter plot of pressure and velocity vectors). From left to right, four columns correspond to phases 0, T/2, 3T/4 and T (T: period corresponding to the frequency) [7]

References

1. Gao Li-Min, Li RuiYu, Zhao Lei, Lin Shi-Yan. Cross-platform impeller machinery high fidelity CFD software (XPLA HF-Turbo CFD). 2020, China Software Copyright, 2020SR1888236.
2. Gao L, Hu X, Tian L, et al. Pressure field measurement on compressor cascade using pressure-sensitive paint. Seoul. Proceedings of ASME Turbo Expo 2016, 2016.
3. Gao L, Li R, Liu B. New Trends and Developments in Metrology – Chapter 7 Optical pressure measurement principle system. Croatia. InTech; 2016.
4. Gao L M, Gao J, Wang H et al. Application of PSP pressure measurement technique in leaf grill blade surface pressure measurement. Journal of Engineering Thermophysics, 2011, 32(3): 411–414.
5. Gao L, Li R, Miao F, et al. Unsteady investigation on tip flow field and rotating stall in counter-rotating axial compressor. Journal of Engineering for Gas Turbines and Power, 2015, 137(7). 1–11.
6. Li R, Gao L, Zhao L, et al. Dominating unsteadiness flow structures in corner separation under high Mach number. AIAA Journal, 2019, 57(7): 2923–2932.
7. Li R, Gao L, Ma C, et al. Corner separation dynamics in a high-speed compressor cascade based on detached-eddy simulation. Aerospace Science and Technology, 2020, 99: 1–11.
8. Yamada K, Furukawa M, Nakano T, et al. Unsteady three-dimensional flow phenomena due to breakdown of tip leakage vortex in a transonic axial compressor rotor. Vienna. Proceedings of ASME Turbo Expo 2004, 2004.
9. Délery, J. Separation in three-dimensional flow. Critical points, separation lines and vortices [DB/CD]. ONERA Information Resources, https://www.onera.fr/sites/default/files/ressources_documentaires/cours-exposes-conf/onera-3d-separation-jean-delery-2011-1a.pdf.
10. Mariappan S, Gardner, A D, Kai, R., et al. Analysis of dynamic stall using dynamic mode decomposition technique. AIAA Journal, 2013, 52(11): 2427–2439.
11. Pan C, Yu D, Wang J. Dynamical mode decomposition of gurney flap wake flow. Theoretical & Applied Mechanics Letters, 2011, 1(1): 1–5.

Chapter 6
Application of DES-Series Method in Compressor Blade Tip Flow Field Capturing

6.1 Introduction

In compressor, tip clearance is an unavoidable structure requirement. The tip leakage flow is one of the most important secondary flows, which has a significant influence on the aerodynamic parameters of the compressor, such as efficiency [1], stability [2] and noise [3, 4]. However, the effective aerodynamic layout and flow control, which relies on accurate prediction of the flow structure of the tip leakage, can effectively enhance the performance. Hence, to attain a more profound and fundamental understanding of the flow field for this enhancement, we aim to utilize the DES-series methodology that balances computational precision with efficiency on the tip leakage flow.

However, unlike the issues encountered in the corner region of the blade, in the tip region, the presence of a tip clearance introduces a significant challenge for the accurate capture and prediction of the strong shear effects between the leakage flow and the mainstream, which is attributed to the abrupt tangential geometrical discontinuity in the flow field. Researches have demonstrated that the traditional DES-series methods, such as DES97 and DDES, are inadequate for predicting shear flows [5–7]. But there is a scarcity of research focusing on the predictive capabilities of the shear interaction between the main flow and leakage flow within a compressor.

Consequently, firstly, we begin with the DDES method's resolution for the detailed flow structure of the tip leakage flow on a cantilevered stator cascade with tip clearance. Subsequently, to delve deeper into the issue of the DDES method's insufficient prediction accuracy in the tip clearance region, a simplified three-dimensional model comprising "multi-wall boundary layers with a narrow gap" was introduced to highlight the characteristics of the blade tip clearance. Based on this, we investigate the reasons for the inaccuracies in the traditional DDES method's prediction from the perspectives of flow detail capture and unsteady turbulence fluctuation, with the LES results as a benchmark. Furthermore, we explored the EDDES method, which is more suitable for tip leakage's prediction and attempted to apply it to the cascade, with the objective of profoundly investigating the loss mechanisms caused by tip leakage.

R. Li et al., *Detached-Eddy Simulation Series Methods for Compressor Internal Flow*,
https://doi.org/10.1007/978-981-96-0255-1_6

Table 6.1 The main geometric parameters of the NPU01 cascade

Parameters	Variable	Value	Unit
No. of blades	–	8	[–]
Blade chord length	C	65	mm
Aspect ratio	h/C	1.523	[–]
Blade solidity	τ	1.73	[–]
Blade stagger angle	δ	1	mm
Tip clearance	β_s	21.27	°
Blade inlet angle	β_{1k}	47.08	°
Blade outlet angle	β_{2k}	– 1.98	°
Camber angle	θ	49.16	°
Design inlet Mach number	Ma_{1d}	0.4	[–]
Design flow incidence	i_d	0	°

6.2 The Accuracy of Tip Leakage Flow Prediction Using the DDES

6.2.1 Compressor Cascade Model with Tip Clearance

A high-loading compressor cantilevered stator cascade, namely NPU01, was selected to investigate the analytical capability of the DDES method for tip leakage flow. The main geometric parameters of the cascade are shown in Table 6.1. The main geometric parameters of the NPU01 cascade, and the schematic diagram is illustrated in Fig. 6.1. The blade height is 99 mm while the tip clearance is 1 mm (1.53%C). The cascade stagger angle χ is 21.7° while the blade inlet angle β_{2k} and blade outlet angle β_{2k} are 47.08° and – 1.98°, respectively, causing the camber angle of the blade is 49.16°. In both experimental and numerical, the incident flow is adjusted by varying the inlet flow angle. The design inlet Mach number is 0.5 with the inlet chord based Reynolds number approximately 7.7 × 10^5 while the design incidence is 0°.

6.2.2 Numerical Method

The focus of this work is to study the predictive performance of the DDES method for tip clearance leakage flow under high incident conditions. The presence of tip clearance causes the flow field structure on the upper and lower walls of the cascade to not be symmetrical about the mid-span height, especially under large separation conditions. The suction surface separation flow, which has a significant impact on the flow at the blade tip region, is also influenced by the corner region of the blade. To ensure the accuracy of the numerical results, we use the entire span of a single passage as the computational model. The height of the first layer of mesh is 2 ×

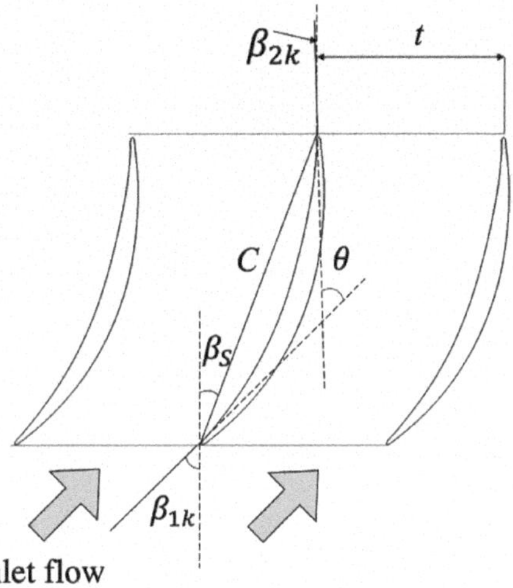

Fig. 6.1 Two-dimensional cascade configuration of NPU01

10^{-6} m, which can ensure $y^+_{wall} < 1$. The mesh generation is completed within NUMECA AutoGrid5™, with a butterfly mesh is used in the tip clearance, while the blade passage and inlet/outlet extensions use an O4H topology. The mesh distribution of the blade-to-blade section (B-B) is shown in Fig. 6.2. As it can be seen, a large number of meshes are arranged around the blade passage, with mesh orthogonality above 43°, as shown in Fig. 6.3. The meshes near the leading edge, trailing edge, and inside the passage are close to cubic, and the streamwise non-dimensional mesh size meets $\Delta x^+ < 110$. Meanwhile, there are 321 meshes in the spanwise direction, and the spanwise non-dimensional mesh size meets $\Delta z^+ < 120$ within the 30% area of the tip and root, causing the total number of meshes is 17.4 million.

The numerical simulation employs the DDES numerical method based on SST, with the solver being the SIMPLE algorithm in FLUENT. The spatial discretization accuracy is of the third order, and the temporal discretization format is a second-order implicit scheme. The physical time step $\Delta t = \overline{\Delta}/Velocity_{\max}$, where $Velocity_{\max}$ is usually taken to be more than 1.5 times the inlet velocity [8], is set here as 1.5 times the inlet velocity; $\overline{\Delta}$ represents the average mesh size, and here it is chosen to be the average mesh size of the core region within the passage at 4.0×10^{-4} m, thus $\Delta t = 1.0 \times 10^{-6}$ s. Regarding the boundary conditions, the inlet is set with total temperature, total pressure, and flow direction, and the outlet is set with static pressure, and the inlet Mach number is altered by adjusting the outlet static pressure. The wall is an

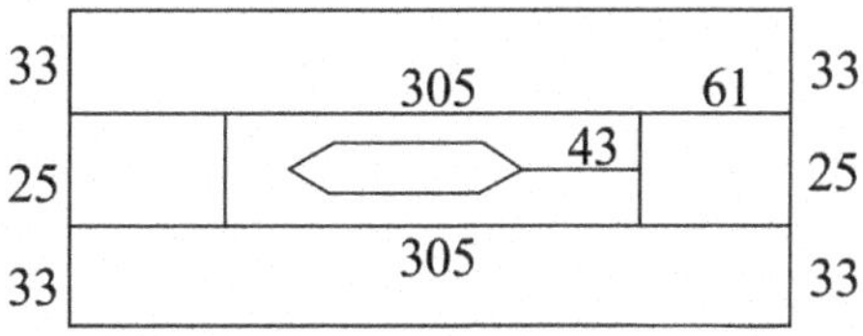

Fig. 6.2 S1 schematic diagram of mesh topology on the S1 flow surface

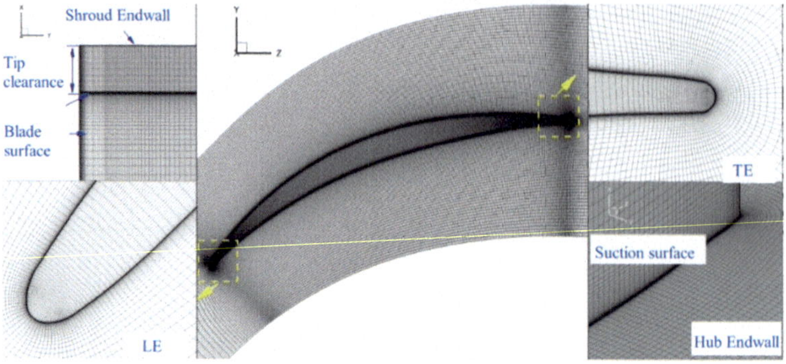

Fig. 6.3 The schematic diagram of the mesh

adiabatic no-slip boundary, and the numerical model passage sides are non-matching periodic boundary conditions. The initial field for the DDES numerical simulation is the result of the converged RANS. To enhance the speed of numerical simulation, an MPI-based parallel method is used, completed on a high-performance computing cluster, equipped with an AMD EPYC 7742 CPU with a main frequency of 2.25 GHz, 256 cores and 512 threads, and this numerical simulation uses a total of 120 cores.

Within the computational domain of the blade tip region, ten static pressure monitoring points were set to assess the convergence of the unsteady numerical simulation, as shown in Fig. 6.4, which illustrates the positions of the ten monitoring points on the S1 flow surface. Convergence is considered to be achieved when the time-averaged results and variance of the monitoring points change by less than 5% over time. Figure 6.5 shows the results of the blade tip static pressure monitoring points; the red dashed line represents the time-averaging results with a window of 5000 steps, advancing with a step size of 500 physical time units, all of which are within 5%, indicating convergence.

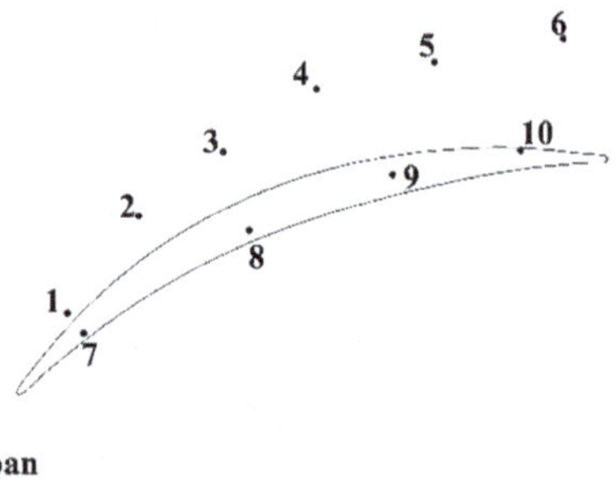

Fig. 6.4 Unsteady static pressure monitoring points in numerical simulation flow field (spanwise located in the middle of the blade tip clearance)

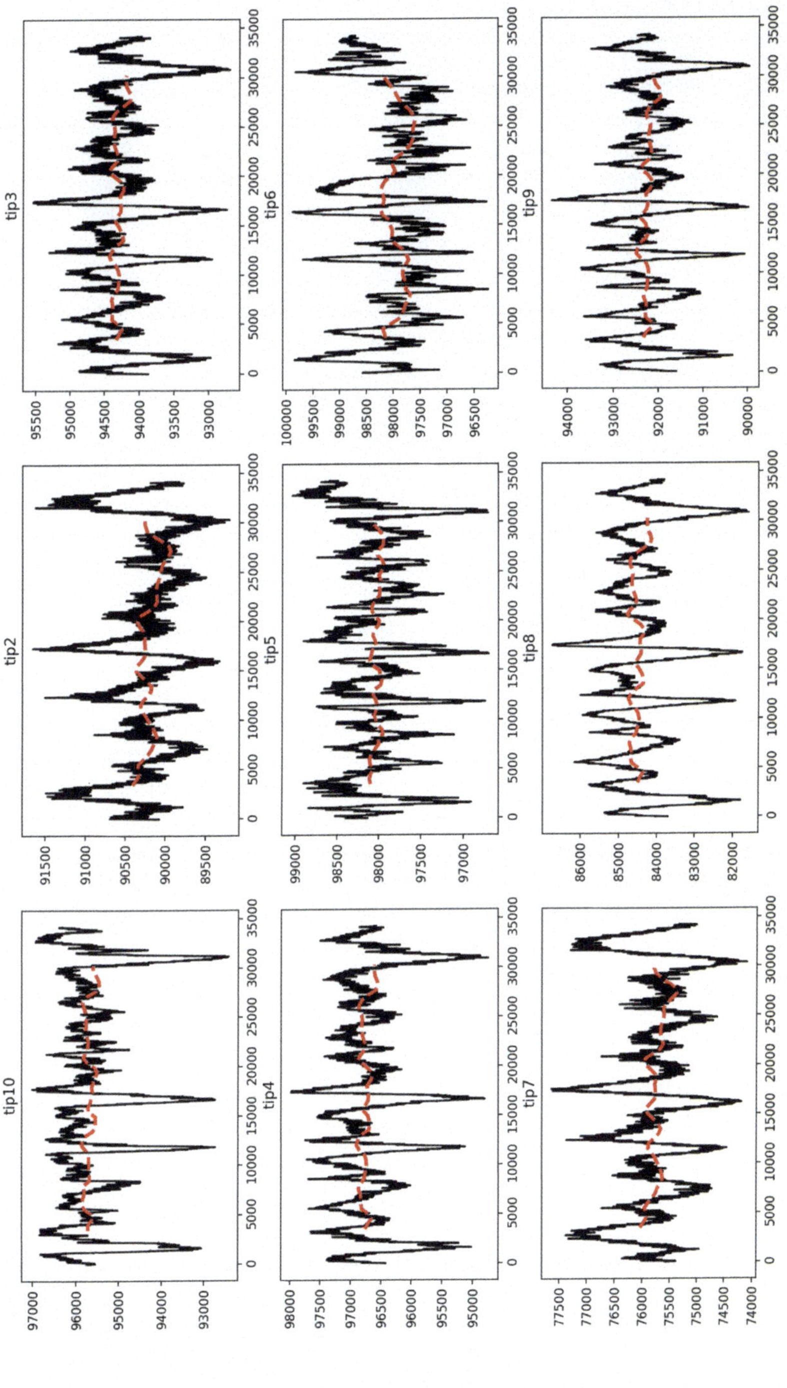

Fig. 6.5 Results of the static pressure monitoring points in the blade tip region

6.2.3 The Detailed Flow Structures in the Tip Region by DDES Method

The unsteady flow field of a compressor is typically composed of the time-averaged flow field, the periodic unsteady flow field formed by the interaction of the rotating and stationary blades, and the turbulent unsteady flow field. Both URANS and RANS methods solve the Reynolds-averaged N-S equations, modeling all the turbulent fluctuations in the flow field through turbulence models. Therefore, although URANS is an unsteady numerical simulation method, it can only resolve the large-scale periodic unsteady flow fields in space and time. For the turbulent fluctuating flow field in the blade tip clearance caused by geometry and adverse pressure gradient in this study, the simulation results of URANS and RANS are similar. Therefore, we compare the numerical results of DDES with those of RANS here.

The transient flow field vortex structures of RANS and DDES are shown in Fig. 6.6, which presents the vortex identification results based on the Q criterion. Qualitatively, it can be seen that since RANS models all fluctuations in the entire flow field (Fig. 6.6a), the turbulent unsteady fluctuations are not resolved. Therefore, the Q criterion results for the RANS simulation results can only reflect information such as the location and extent of separation (as shown in Fig. 6.6a). The results in Fig. 6.6b show the results of the DDES method, and it can be seen that there is a significant improvement in the turbulent resolution and the capture of the vortex structures for the mid-blade and corner separation regions, especially below 75% of the blade span. The spatial coherent structure such as corner separation vortices, suction surface separation, and concentrated shedding vortices have been greatly enhanced compared to the RANS method. Unfortunately, the DDES method still falls short in resolving the tip region, and its resolution and capture of vortex structures are similar to the RANS results, especially in the parts before 75% of the chord length. Even under the extreme condition of a 6° incident angle, it still fails to capture the dynamic characteristics of the vortex structures, including the fluctuation, breakdown of the tip leakage vortex (TLV), and the mixing between different vortex structures. Therefore, we need to explore a adapt version of the DDES method that is more suitable for the leakage flow at the tip region.

6.3 Simplified Model of Tip Clearance Leakage Flow

Through the aforementioned studies, it can be observed that even when the grid size at the tip and root is consistent and the same method, the DDES method, was used, the turbulent resolution and capture of the vortex structures at the tip region remains relatively low. This indicates that the unique geometric characteristics of the tip clearance are the core cause of the inaccuracy in the DDES method's predictions. To investigate a DDES method more suitable for the tip clearance from a mechanistic

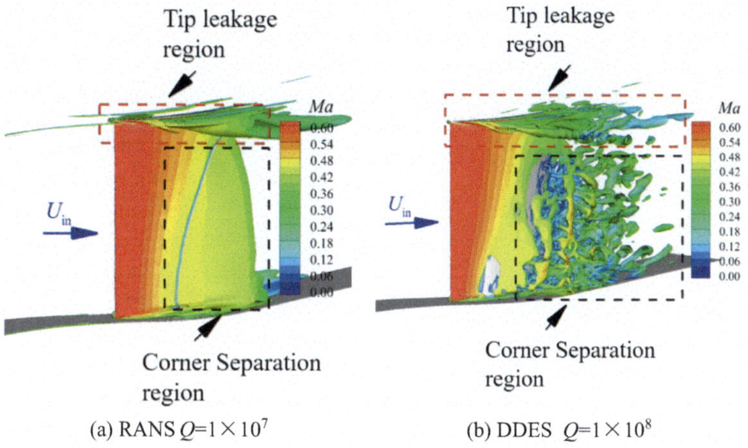

Fig. 6.6 The vortex structure diagram of the full-span flow field within the passage based on the Q criterion

perspective, we first established a simplified model that can represent the unique aerodynamic characteristics of the blade tip, which include "multiple boundary layers with shearing effect between mainstream and leakage flows". Based on this, we explore the influence mechanism of the DDES method on the accuracy of the prediction of the tip clearance flow from the detailed vortex structure and unsteady dynamic characteristic.

6.3.1 Model Establishment

To emphasize the special geometry of the tip leakage, i.e., the "multi-wall + narrow tip gap" structure and the corresponding aerodynamic characteristics, the simplified model comprised a square passage and a tip gap was established as shown in Fig. 6.7. The square passage is a simplification of the cascade passage consisting of a top wall, two side walls, and a symmetry plane at the bottom. Additionally, the tip gap is a simplified tip clearance structure composed of upper and lower solid walls connected to the square passage. According to the model, the mainstream along the X direction in the square interacts with the leakage flow from the tip gap with an angle of α, which forms the TLV.

The sizes of the simplified model geometry abstracts from a compressor cascade are shown in Table 6.2, where L_y and L_z are the width and height of the square passage, corresponding to the pitch of the cascade and blade height, respectively. L_{tip}, L_t, and τ are the size of the tip gap, corresponding to the axial chord length of the blade, the maximum thickness of the blade, and the tip clearance size, respectively. α is the angle between the tip gap and the square passage, representing the angle between the leakage flow and the mainstream. L_0 is the distance between the start position

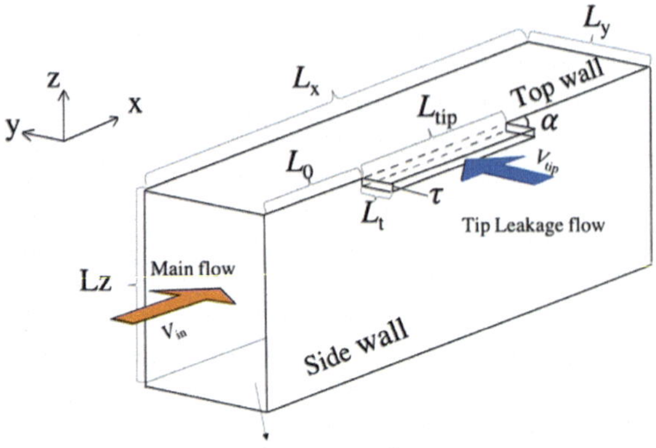

Fig. 6.7 The diagram of the simplified model

Table 6.2 The parameters of the simplified model

Parameters	Value
L_z	0.67 h
L_y	0.63 h
L_0	h
L_x	4 h
L_{tip}	h
L_t	0.083 h
τ	0.003 h
α	90°

of the tip gap and inlet, corresponding to the distance between leading edge of the blade and the inlet. L_x is the axial length of the whole calculation domain. The size of the tip gap channel in this paper is $L_{tip} = h = 0.06\,\text{m}$, and the specific values of other parameters are given in Table 6.2.

6.3.2 *Numerical Methods*

The wall-adapted local eddy-viscosity model large-eddy simulation (WALE LES) and the DDES methods are selected to solve the three-dimensional transient N-S equations, respectively, and based on the CFX solver, the subgrid-scale viscosity coefficient (SGS) μ_{sgs} is obtained by solving the Smagorinsky subgrid model with the WALE LES method. Compared to the original Smagorinsky model, the WALE LES method (without ambiguity, the following is called LES) ensures the eddy-viscosity coefficient in the viscous sublayer is close to zero and captures the transition from laminar to turbulent flow at the near-wall region [9]. The DDES method used in this paper is based on the SST model [10]. The mixed-length l_{hyb} is used instead of l_{t} in the original SST $k-\omega$ by introducing the delay function F_{DES} based on the mixed function F_2, see Eqs. (6.1)–(6.5)

$$F_{DES} = \max\left(\frac{L_t}{C_{DES}\Delta}(1-F_2), 1\right) \tag{6.1}$$

$$L_t = \frac{\sqrt{k}}{\beta^* \omega} \tag{6.2}$$

$$\Delta = \max(\Delta_x, \Delta_y, \Delta_z) \tag{6.3}$$

$$F_2 = \tanh(\arg_2^2) \tag{6.4}$$

$$\arg_2 = \max\left(2\frac{\sqrt{k}}{\beta^* \omega d}, \frac{500\nu}{d^2\omega}\right) \tag{6.5}$$

The delay in the transformation from RANS ($F_2 = 1$) to LES ($F_2 = 0$) is achieved through F_2, which can avoid the problem of GIS [10]. In terms of the discrete format, we use the central difference scheme as the discrete spatial convection, whereas the second-order backward Euler is used for the time discrete convection. Regarding the boundary conditions, although we set the average pressure at the outlet, the total temperature and pressure are set at the inlet. The velocity of the mainstream and jet flow are controlled by adjusting the total pressure at the inlet of the square and tip gap (mainstream's Mach number: 0.4). Anyway, the Mach number of the leakage is set to 0.4 to ensure the calculation's convergence stability and avoid the backflow at the tip gap inlet.

As shown in Fig. 6.8, the calculation domain of the simplified model comprised two H-type domains, the square passage and the tip gap with $301 \times 201 \times 205$ ($X \times Y \times Z$) and $201 \times 73 \times 61$ ($X \times Y \times Z$) cells, respectively. Notably, the grid here in the red box is refined at the position of the leakage and mainstream's shear layer; that is, the focus region, to meet the requirements of the WALE LES method that is $y^+ < 1$ near the solid walls and $x^+ < 40$ and $z^+ < 30$ at the focus region in figure, resulting in a total grid number of 15.54 million. Contrastingly, the same grid is used in the DDES calculation to eliminate grid-based influence. The physical time step $\Delta t_{physics}$ is set to 1×10^{-6} s, ensuring the Courant number is less than 1.

The IQ_v is a variable used to evaluate the proportion of the resolved TKE to the total TKE. A higher IQ_v implies a larger percentage of resolved turbulence [11]. As is shown in Fig. 6.9, for the resolved turbulence of the leakage vortex, the IQ_v of the LES method is greater than 0.8, whereas the IQ_v of the DDES method is less than 0.8, indicating that the DDES method does not reach the LES level needed for the resolved turbulence of the generation and development downstream of the leakage vortex.

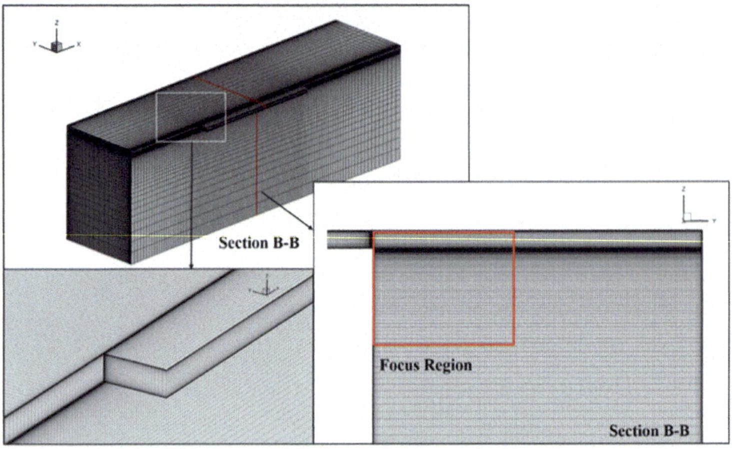

Fig. 6.8 The grid of the simplified model

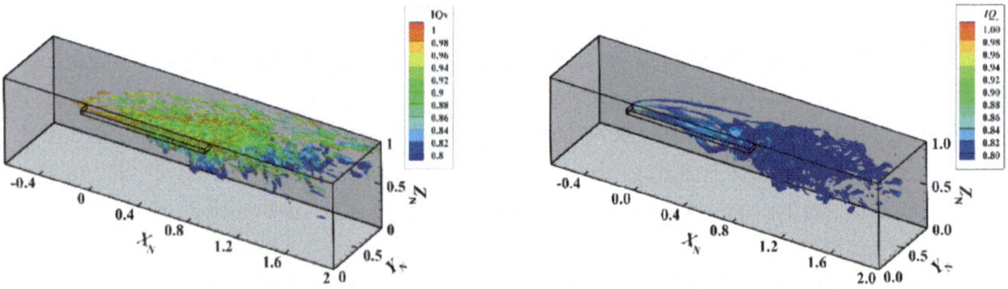

(a) LES (b) DDES

Fig. 6.9 The iso-surface of $Liutex_{\mathrm{mag}} = 30{,}000$ (contoured by IQ_v)

6.3.3 Accuracy Analysis

Comparison of the Resolution of the Turbulence

To verify the LES and DDES resolution in the tip field, Fig. 6.10 shows the power spectral density of the pressure fluctuation at five representative monitor points. Figure 6.10a demonstrates the positions of the detection points, where point 1 is at TLV's starting position, and point 2–point 5 are arranged along the leakage vortex core's trajectory to ensure that the numerical monitoring point remains in the fluctuating area caused by the leakage flow. The vertical coordinate E denotes the energy density at the corresponding frequency. In this study, the parametric autoregressive (AR) method based on the Burg method [12] is used for data resulting from the unsteady results, which has a good quality spectral estimate even for short data

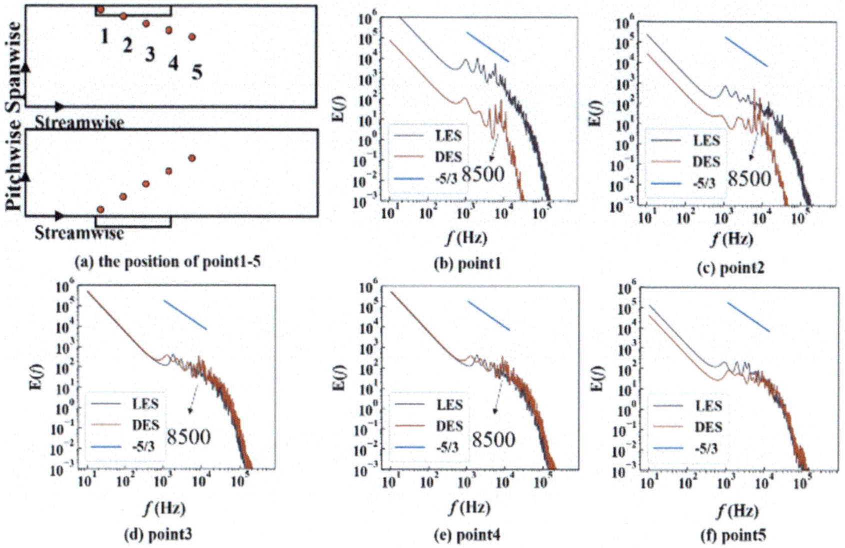

Fig. 6.10 Power spectral density distribution of the static pressure at monitoring points

records. With a slope of – 5/3 in the inertial subrange (10^3–10^4 Hz), the power spectral densities of the five monitoring points based on LES were roughly consistent, proving the reliability of the LES results.

Conversely, the DDES results show that the power spectral density of the two near-wall monitoring points (i.e., point 1 and point 2) at the beginning of the leakage vortex did not match the slope of – 5/3, indicating that the TKE captured by DDES at these two positions was much smaller than that of LES. Investigations also reveal that a different high-energy region from the LES results exists near the 8500 Hz frequency band, showing that the resolved TKE at point 1 and point 2 does not reach the LES level. At point 3 and point 4, however, although the power spectral density results obtained by the two methods are relatively close, slight differences still exist at 8500 Hz, indicating that the DDES method has a local high-fluctuation energy region. Then, for point 5, the farthest point from the wall, the power spectral density of the DDES and LES results are very similar.

Comparison of the Detailed Flow Structure

The transient vortex structure of the simplified model of the LES and DDES results are as shown in Fig. 6.11 and the iso-surface is $Liutex_{mag} = 30{,}000$, which is contoured by the static pressure. In the LES result, the leakage flow ejects via the tip gap and interacts with the mainstream, forming spiral-like swirling flow structures with the minimum pressure, namely, the TLV. It generates at the starting position of the tip gap ($X_N = 0$) and develops in the direction of $Y+$ and $Z-$, surrounded by the secondary leakage vortex (STLV). The red box in Fig. 6.11 shows that the STLV generated at the position of $X_N = 0.2$ and enters the square passage in multiple strands, gradually

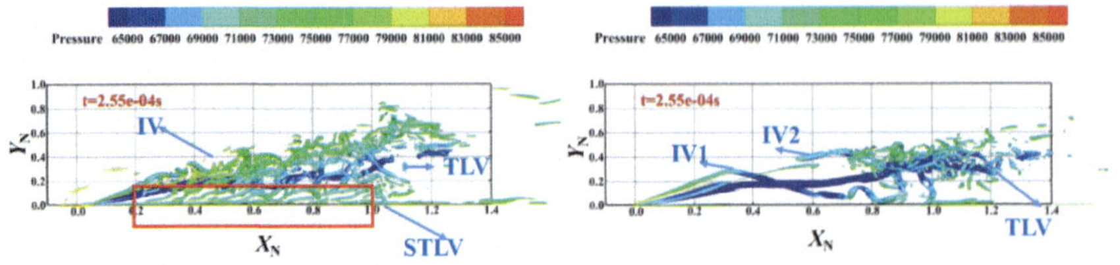

(a) LES (b) DDES

Fig. 6.11 The transient vortex structure of the simplified model (iso-surface $Liutex_{mag} = 30{,}000$)

merging into the TLV. Besides, there is an induced vortex (IV) induced by the shear effect of the leakage flow and the boundary layer of the upper endwall outside ($Y+$) the TLV. Furthermore, the vortex core of the IV shakes and twists, and then breaks up into large-scale turbulence at the position of $X_N = 0.4$.

Upon comparing the DDES transient flow field structures (Fig. 6.11b) with the LES structures (Fig. 6.11a), it is observed that DDES has captured the TLV and the IV. However, there are two notable differences:

1. Capture of the STLV: The LES method detects the initiation of the STLV at the position $X_N = 0.2$, which then divides into several streams entering the passage and gradually merging into the TLV. This observation aligns well with the physical flow field. In contrast, DDES fails to capture the STLV, meaning that DDES does not resolve the fluctuations of the STLV. As a result, there is a discrepancy in depicting the impact of the STLV fluctuations on the TLV's dynamics.
2. Capture of the IV: The IV, as captured by the LES method, undergoes expansion and fragmentation at the position $X_N = 0.4$, interacting and interfering with the TLV. In the case of DDES, the IV is divided into two branches. IV1 wraps around the underside of the TLV at $X_N = 0.35$ and never undergoes breakdown. IV2 remains outside the TLV and begins to expand at $X_N = 0.72$, interacting with the TLV. This indicates that DDES does not resolve the formation of the STLV upstream of the leakage and that the IV do not breakdown and dissipate in a timely manner in the mid to downstream regions of the leakage.

The vortex structures captured by the two methods exhibit distinct differences, which can lead to variations in the prediction of velocity gradients and shear effects. Consequently, it can be inferred that these discrepancies may impact the main fluctuation frequencies and corresponding amplitude of the flow field.

6.4 DES-Series Methods for Capturing Tip Clearance Flow

6.4.1 The Reason for the Inaccurate Prediction in the Detailed Structures

Through the aforementioned studies, it can be seen that the vortex structures captured by the two methods exhibit distinct differences, which can lead to variations in the prediction of velocity gradients and shear effects. Consequently, it can be inferred that these discrepancies may impact the main fluctuation frequencies and corresponding amplitude of the flow field. Furthermore, the main fluctuations within a flow field are significant factors affecting the time-averaged flow field, and since losses in the flow field typically arise from energy exchanges between quasi-coherent structures, these main fluctuations are often the primary source of losses. In this section, the SPOD method is employed to decompose and reconstruct the main frequency flow field, investigating the ability of the DDES method to capture the main frequency unsteady fluctuations in the flow field.

Firstly, we extracted the static pressure data from the blue region in Fig. 6.12 for the SPOD decomposition, the flow direction, tangential direction, and spanwise range are $X_N \in [0, 1.2]$, $Y_N \in [-0.013, 0.6]$, and $Z_N \in [0.5, 1.0]$, respectively. The input data are the flow field snapshots at 2048 moments, each of which contains static pressure fluctuation data at 6.82 million spatial points. Anyway, the physical time interval between the two snapshots is 5×10^{-6} s, namely $5 \times \Delta t_{physical}$. Moreover, the data are divided into eight block matrices (i.e., $N_b = 8$). The snapshot number N_f in each block matrix is 1024, and the overlapping snapshot number N_o between two blocks is 880. The Hanning window is selected as the window function. The weight function W of each cell is based on the normalized volume of the local grid $W = V_i/\overline{V}$. Finally, we complete the calculation and reconstruction of the SPOD decomposition using a 2T memory computing node on the Tianhe II server.

Figure 6.13 shows the energy spectrum of the first five modes obtained by SPOD decomposition. It can be seen that since the energy ratio of the first-order mode is significantly higher than the others and the energy ratio of the fifth-order mode is tiny, the modes after mode 5 are negligible here. Overall, from the energy spectrum

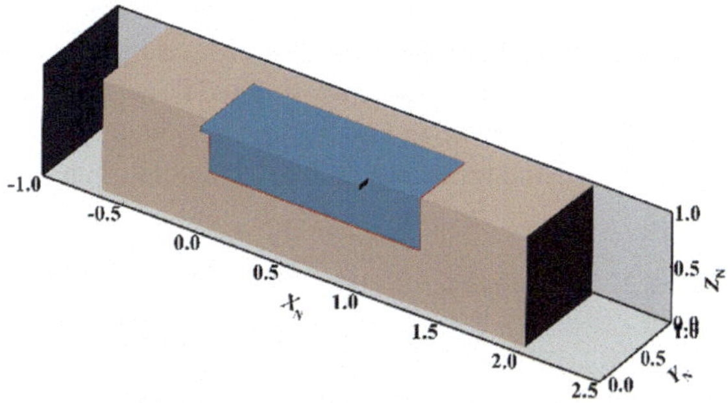

Fig. 6.12 The three-dimensional data selection range of the SPOD decomposition

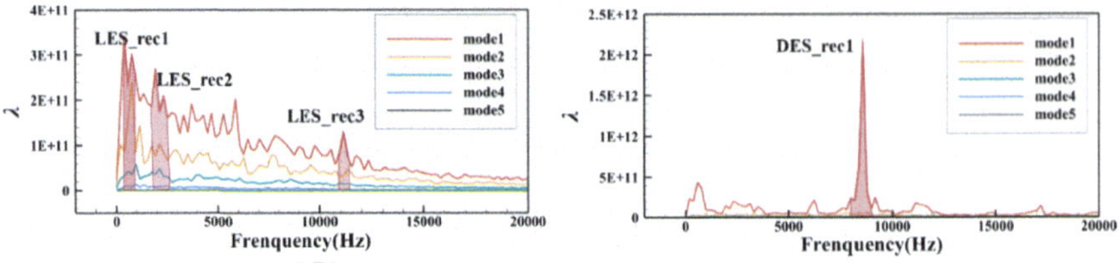

Fig. 6.13 The energy spectrum of the first five modes of SPOD decomposition

perspective, the LES results exhibit the following characteristics: (1) mode 1 and mode 2 occupy the majority of the fluctuation energy and the energy distribution of both are close, being high in low frequency and low in high frequency; (2) the energy frequency band is rich, possessing both multi-band and wide spectrum characteristics. Conversely, the DDES result is quite different: (1) the fluctuation energy is primarily concentrated in mode 1, with a higher energy distribution at a specific frequency (8000–9000 Hz); (2) the main frequency band occupies the majority of fluctuation energy, whereas the high-frequency energy band is narrow. Compared with the rich frequency distribution of LES, the flow field simulated by DDES has a significant dominant frequency; that is, high-energy fluctuations are captured within 8000–9000 Hz.

Since the coherent structure with a high fluctuation energy has the most significant influence on the flow in the unsteady flow field, we selected high-energy frequency bands with the first-order modes from the LES and DDES results for flow field reconstruction. Table 6.3 shows the selected frequency bands. The reconstruction result is the superposition of the flow field at a selected frequency band in the first-order mode.

Figure 6.14 shows the reconstructed flow field at a specific frequency obtained by the LES and DDES methods, respectively. The red and green iso-surfaces are the pressure fluctuation iso-surfaces in the reconstructed flow fields corresponding to the specific frequency. They indicate the main fluctuation position. Due to the fluctuation energy differences corresponding to each frequency band, we varied the values of the iso-surface to show the core fluctuation position in the reconstruction results of the specific, as depicted in the figure. The blue iso-surface in each figure is $Liutex_{\text{mag}} =$ 10,000, representing the vortex structure of a transient flow field. It could intuitively reflect the physical phenomena corresponding to the main fluctuation.

For the LES results, the reconstructed flow field LES_rec1 with a low frequency of 700–1000 Hz is shown in Fig. 6.14a. The fluctuation amplitude of this frequency band is the largest, and the major source of fluctuation is the shake of the TLV core. In the

Table 6.3 The reconstructed frequency band

LES			DDES
LES_rec1	LES_rec2	LES_rec3	DDES_rec1
700–1000	2000–3000	10,500–11,500	8000–9000

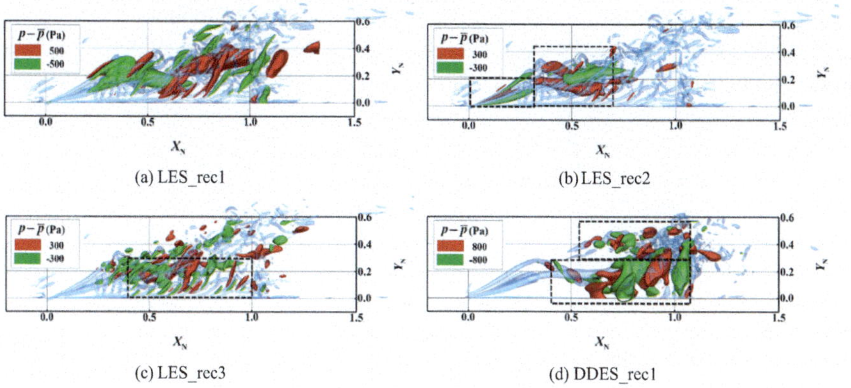

Fig. 6.14 The reconstruction field of the specific frequency band based on LES and DDES results

reconstructed flow field of LES_rec2 (Fig. 6.12b), the pressure fluctuation appears in the upstream region of the TLV and the IV in the space of $X_N \in [0, 0.32]$, and the downstream region of the interaction between the two vortices of $X_N \in [0.32, 0.70]$. In the pressure fluctuation flow field corresponding to the high frequency of 10,500–11,500 Hz, as shown in the black box in Fig. 6.14c, the fluctuation position is more complicated, which is the mixing result of the generation of the STLV, the breakdown of the TLV and IV. On the contrary, from the reconstruction results of DDES in Fig. 6.14d, the fluctuation of the flow field comes from the breakdown of the two IV branches and the interference with the TLV.

Overall, the above results indicate that the main fluctuation sources of the LES flow field from large to small are as follows: the TLV core shakes (located along the leakage flow), the interference between IV and TLV (mainly located upstream of the leakage), and the influence of STLV (mainly located at the middle and downstream of the leakage). Additionally, we observed that the major fluctuations captured in the DDES results only come from the IV, comprising the breaking of the IV1 and IV2, and their interference mixing with the TLV (downstream of the leakage).

In order to reveal the mechanism that DDES's failure in capturing the detailed unsteady flow structures, the $X_N = 0.67$ cross-section was extracted for SPOD decomposition and to reconstruct the main fluctuation flow field. The $q = [p', V_{x'}, V_{y'}, V_{z'}, \Delta s']$ is chosen as the fluctuation snapshot, and the weight function is shown in Eq. (6.6), following the principle of total disturbance energy [13] that the energy of fluctuation contains the turbulent energy, the entropy potential, and the pressure potential. The input data are snapshots of the flow field at 2048 moments, each containing static pressure fluctuation data at 46,000 spatial points. Other relevant settings are consistent with three-dimension SPOD above. The multi-parameter fluctuation SPOD energy spectrum of the $X_N = 0.67$ slice is shown in Fig. 6.15. The 10,600–11,500 and 8000–9000 Hz frequency bands in the first-order mode representing the main fluctuation in LES and DDES are reconstructed, respectively, and the reconstruction results are shown in Fig. 6.16. The contour is pressure fluctuation, and the vector is the projection of this section's velocity fluctuation. At the same time, the partially enlarged view in Fig. 6.16 gives the tangential velocity gradient in the spanwise direction.

$$
\begin{aligned}
W &= \frac{1}{2}\int_V \left(\frac{R\overline{T}}{\overline{\rho}}\rho'^2 + \frac{R\overline{\rho}}{(\gamma-1)\overline{T}}T'^2 + \overline{\rho}u_i'^2\right)dV \\
&= \frac{1}{2}\int_V \left(\frac{1}{\gamma\overline{p}}p'^2 + \frac{(\gamma-1)\overline{p}}{\gamma R^2}\Delta s'^2 + \overline{\rho}u_i'^2\right)dV
\end{aligned}
\tag{6.6}
$$

It is evident from Fig. 6.16 that noticeable tangential velocity gradients are generated when the secondary leakage flow ejects from the tip gap and shears with the mainstream and consequently forms the Kelvin–Helmholtz (K-H) instability in the area enclosed by the black box. Driven by K-H instability, the shear layer generates strong unsteady fluctuations. Subsequently, it forms STLV along the tangential direction, corresponding to the black dotted arrow in Fig. 6.16a, and finally confluences

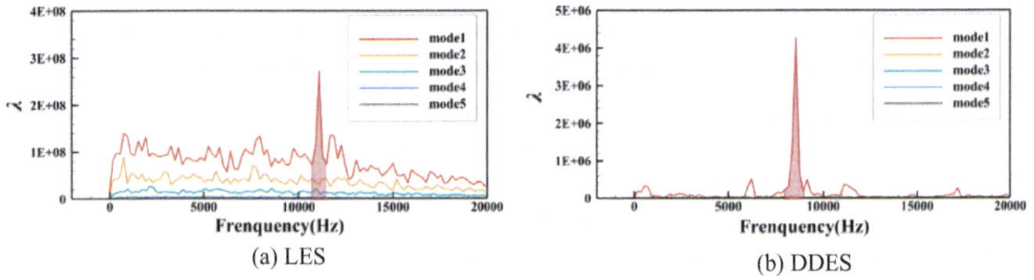

Fig. 6.15 The multi-parameter fluctuation SPOD energy spectrum of the $X_N = 0.67$ section

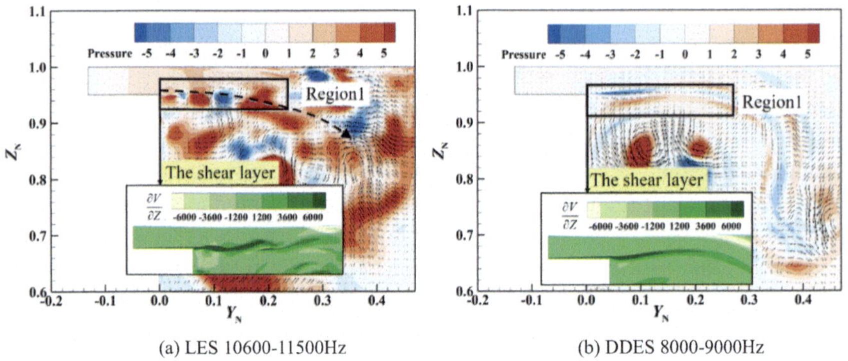

Fig. 6.16 Flow field reconstruction of high-energy band based on multi-parameters ($X_N = 0.67$ section)

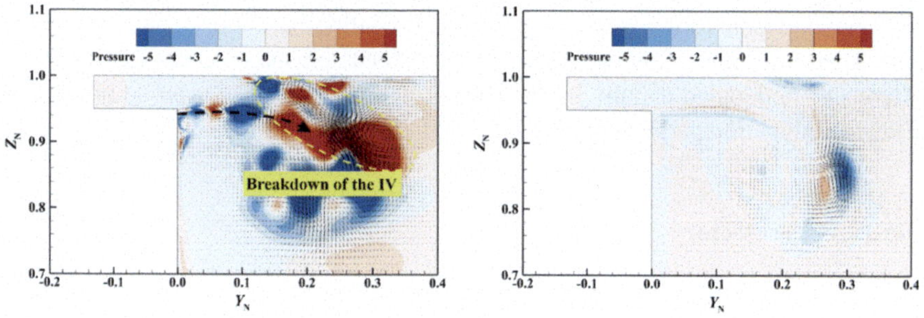

(a) LES (b) DES

Fig. 6.17 The reconstruction results of the high-energy band at $X_N = 0.34$

into the TLV. This result is similar to that shown in the high-energy fluctuation reconstruction results of SPOD in Fig. 6.14c, indicating that the STLV structure comes from K-H instability.

To explore the reason why the IV in the DDES result is unbroken at the position $X_N = 0.4$, the reconstruction results of the high-energy bands based on SPOD multi-parameter decomposition at the $X_N = 0.34$ slice (the upstream position of the breakdown of IV) are given in Fig. 6.17. It can be intuitively seen that the pressure and velocity fluctuation captured by LES is obviously larger than that of the DES method. The yellow dotted area corresponds to the IV range. It is obvious that IV breaks under the impact of the secondary leakage fluctuation, corresponding to the black dotted arrow in Fig. 6.17a. On the contrary, the fluctuation of the secondary leakage flow is suppressed in the DDES results (Fig. 6.17b), and accordingly, the insufficient TKE causes the IV not to break up. Subsequently, IV develops downstream into IV1 and IV2. In other words, the fluctuation of STLV is the main reason for the breakdown of IV.

In sum, In the DDES simulation, the K-H instability is suppressed in the Region 1 where the leakage flow ejects out of the gap and shears with the mainstream. and the unsteady fluctuation cannot be correctly analyzed. This results in the inaccurate capture of the STLV generation and IV development, namely, the deviations in the DDES method's prediction of detailed structured. Therefore, correctly capturing K-H instability is the key to improving the accuracy of tip clearance flow field simulation.

6.4.2 *Impact of Dominant Flow Structures on Time-Averaged Parameters*

The detailed flow structures have a significant impact on the aerodynamic parameters. For instance, the shear and squeezing actions between vortex structures mentioned above can affect the development trajectory and breakdown of the tip leakage vortex, thereby influencing aerodynamic stability; the mixing, interference, and momentum exchange between vortex structures can affect the distribution of losses in the flow field. Therefore, the following discussion will focus on the mechanism which the differences in detailed capture by the DDES method affect the accurate prediction of key aerodynamic parameters.

The TLV's Trajectory

As shown in Fig. 6.18a, the trajectory of the TLV in the numerical flow field of LES is affected by two vortex structures in the tangential direction, namely the IV and STLV. The STLV acts on the TLV in the Y+ direction through viscous shear. The vector of the IV is opposite to the vector direction of the TLV, rotating clockwise along the flow direction. Therefore, there is a Y− direction force on the leakage vortex in the tangential direction. The leakage vortex trajectory is the result of the combined action of these two forces in different directions. From the SPOD decomposition results, the effect of the STLV is more notable after $X_N = 0.4$ (Fig. 6.14c). The effect of the IV on the leakage vortex, before and after the vortex breakdown, is mainly in the range of $X_N \in [0, 0.7]$, as shown in Fig. 6.14b. It means that the trajectory of TLV is only affected by the force of the STLV in the Y+ direction after $X_N > 0.7$. The SPOD decomposition results (Fig. 6.14d) show the numerical field of DDES, with the trajectory of the TLV being mainly affected by the two IV branches. As shown in Fig. 6.18b, although the vortex vector directions of the two branches are the same, the directions of the force acting on the TLV are different due to their different spatial positions. The force of IV2 on the outer side of the TLV is along the Y− direction. For IV1, in the range of $X_N \in [0.6, 0.9]$, because it is located inside the leakage vortex, its tangential force on the leakage vortex is along the Y+ direction, and the leakage vortex trajectory is affected by two IV branches. After $X_N > 0.9$, IV1 has turned to the upper side of the leakage vortex and broken, providing an obvious force along the Z-direction.

In summary, within $X_N \in [0, 0.9]$, the leakage vortex in the LES and DDES flow fields is affected by the forces in two different directions in the tangential direction. Although the sources are different, the trajectory of the TLV is relatively close from the time-average point of view. However, after $X_N > 0.9$, because the leakage vortex in the LES flow field is only affected by the STLV of the Y+ force, the trajectory is more inclined to Y+, whereas the DDES is only affected by the IV2 with Y− direction

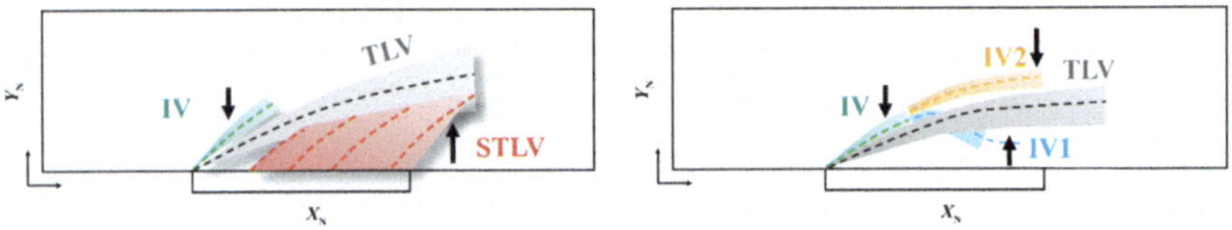

(a) LES (b) DES

Fig. 6.18 The diagram of the force direction on the TLV

force. In the spanwise direction, the IV1 turns to the upper side of the leakage vortex at $X_N > 0.9$, and the vortex core expands and squeezes the TLV, resulting in the TLV trajectory more biased towards Z− direction based on DES. It can be concluded that the difference in the prediction of the STLV and IV development is the main reason for differing downstream leakage vortex trajectory predictions.

The Loss of the Tip Region

In the simplified model, the source of the loss is relatively straightforward because no complex physical phenomena, such as separation caused by shock waves, wakes, and adverse pressure gradients, exist. Theoretically, the flow loss (non-near wall) in the simplified model mainly comes from flow shear and vortex mixing. The SPOD decomposition results show that the vortex interaction mainly appears in the $X_N > 0.4$ region rather than $X_N \in [0, 0.4]$, which indicates that the loss in $X_N \in [0, 0.4]$ is dominated by flow shear. Therefore, we mainly analyze the causes of the loss differences in the range of $X_N > 0.4$.

Furthermore, according to the SPOD decomposition results (Fig. 6.14b) from the flow field simulated by LES, the major fluctuation appears in $X_N \in [0.4, 0.7]$ where the breakdown of IV occurs, which implies that the vortex breakdown is the main source of the loss. In contrast, instead of breaking, the IV of the DDES flow field develops into two branches IV1 and IV2, causing the main fluctuations at Region 2 and Region 3 in Fig. 6.19, which dominates in the loss of this range. Subsequently, due to the breakdown of the IV1 and IV2 in the DES flow field, the range of $X_N > 0.9$ becomes the main fluctuation region according to the SPOD decomposition results (Fig. 6.14d), whereas the fluctuation in the LES flow field here has stabilized after mixing. In consequence, the loss predicted by DDES is small in $X_N \in [0.4, 0.7]$ and large in $X_N > 0.9$. Therefore, it can be concluded that the difference in the capture of the IV development is the main reason for loss prediction differences.

In sum, due to the influence of the K-H instability's suppression, resulting in the inaccurate capture of the STLV generation and IV development, both of which affect

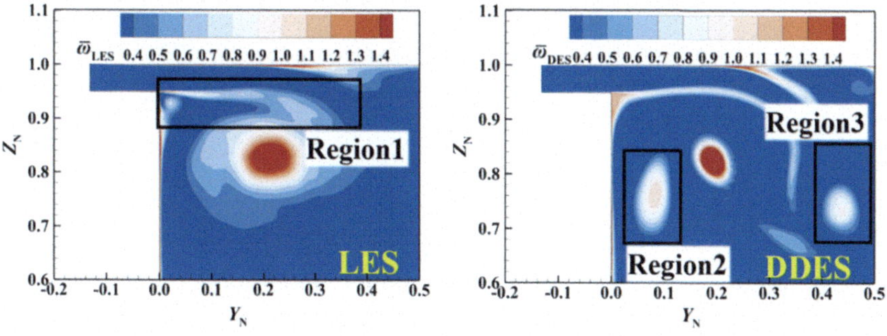

Fig. 6.19 The time-averaged loss based on LES and DES methods at $X_N = 0.67$ section

the trajectory of the vortex core. Furthermore, the diversity of IV development leads to the inaccurate prediction of the main fluctuation sources in the flow field, causing inaccurate total pressure loss predictions in the DES method. Therefore, correctly capturing K-H instability is the key to improving the accuracy of tip clearance flow field simulation.

6.4.3 Accuracy Analysis of the EDDES Based on Simplified Model

The EDDES method, which was invented to enhance the capture of K-H instability at the initial shear location, is highly anticipated in addressing the inaccuracy issues of DES-series methods in tip predictions. To validate the application effectiveness of the EDDES method in the tip region, numerical simulations were conducted on the simplified model by the EDDES methods.

The open-source CFD software SU2 (Stanford University Unstructured) [14] from Stanford University was utilized to perform numerical simulations on the simplified model, employing EDDES for the turbulence method. In terms of discretization, the convective terms were spatially discretized using a second-order central difference scheme, and the temporal advancement was based on a dual-time stepping method with a second-order backward Euler approach. The boundary conditions were set in accordance with the numerical simulation configurations described in Sect. 6.3.2. The mesh employed an OH-structured grid, as shown in Fig. 6.20, to enhance computational convergence within the SU2 software. The total number of grid cells is 13.4 million. In the region of interest, which is the main fluctuation region of the leakage flow (based on Sect. 6.3.3), the grid size is consistent with that used in Sect. 6.3.2.

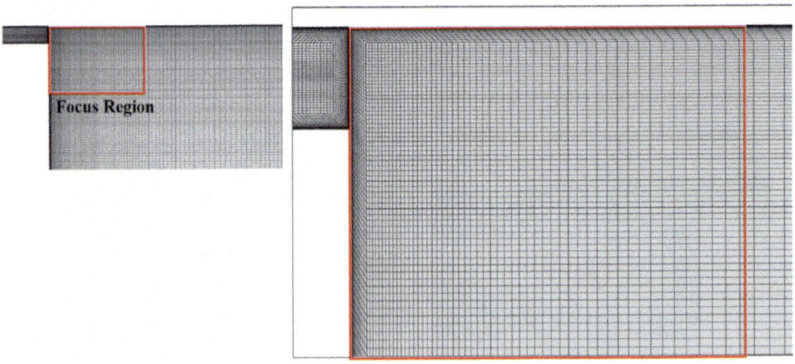

Fig. 6.20 The OH-type grid of the "three-dimensional wall + narrow gap structure" simplified model

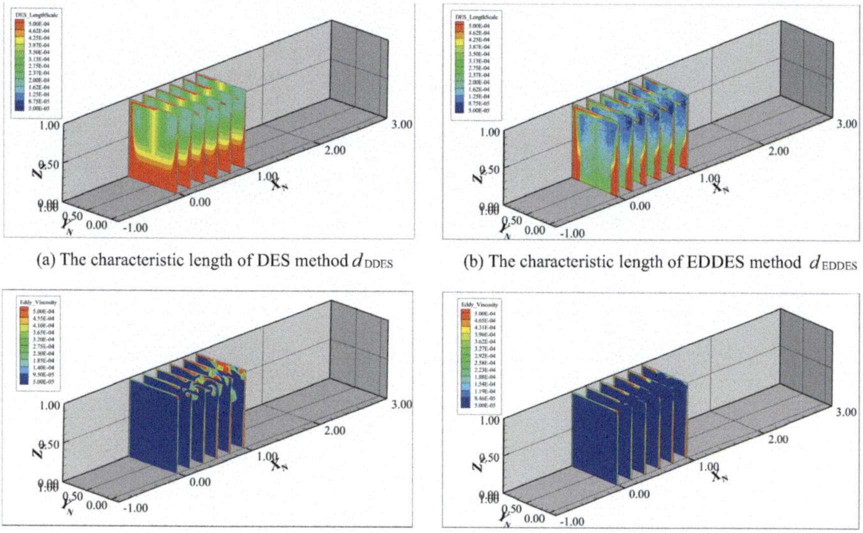

(a) The characteristic length of DES method d_{DDES}

(b) The characteristic length of EDDES method d_{EDDES}

(c) The eddy viscosity coefficients of DDES method

(d) The eddy viscosity coefficients of EDDES method

Fig. 6.21 The characteristic lengths and eddy viscosity coefficients in DDES and EDDES methods

As shown in Fig. 6.21a, b, in the initial shear region, the characteristic length $\tilde{d}_{\mathrm{EDDES}}$ in the EDDES method is significantly reduced compared to that in the DES method, especially in the area near the tip leakage vortex. Consequently, the eddy viscosity coefficient in the EDDES method for the initial shear region is markedly lower than that in the DES method.

Comparing the EDDES (Fig. 6.22c), with LES (Fig. 6.22a), and the DDES transient flow field structure (Fig. 6.22b), it can be observed that EDDES captures a richer vortex structure than DDES method. The following examination focuses on the inaccuracies of the DDES method in capturing two key features of the flow field—the generation of the STLV and the breakdown of the IV—and evaluates the improvements made by EDDES.

(1) Capture of the STLV: The EDDES method begins to capture the generation of the secondary leakage vortex after $X_{\mathrm{N}} = 0.75$, which is an improvement over the DDES method, which fails to capture the STLV throughout the entire flow process. However, there is still a gap compared to the LES method, which captures the generation of the STLV at $X_{\mathrm{N}} = 0.2$.
(2) Prediction of the IV breakdown: In the results of the EDDES method, the induced vortex has not breakdown by $X_{\mathrm{N}} = 0.4$ and begins to move in the Y-direction, wrapping from the underside of the leakage vortex to the inner side of the leakage vortex, which is similar to the DES method.

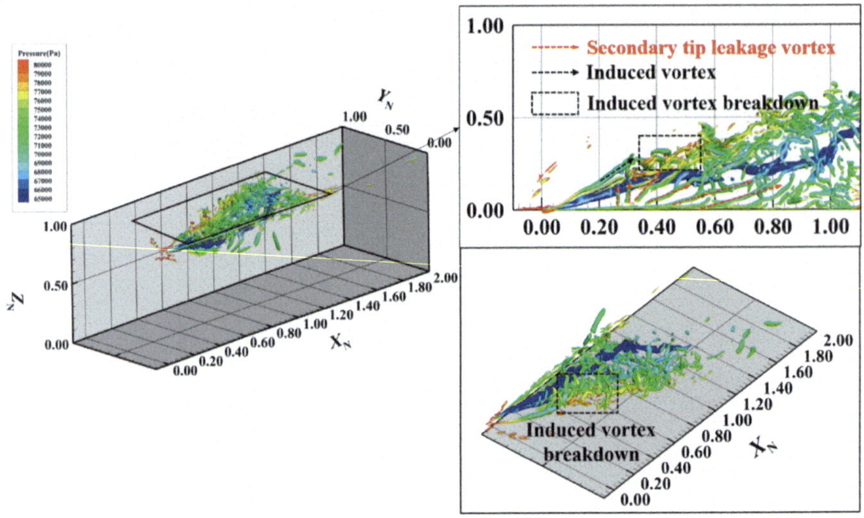

(a) LES

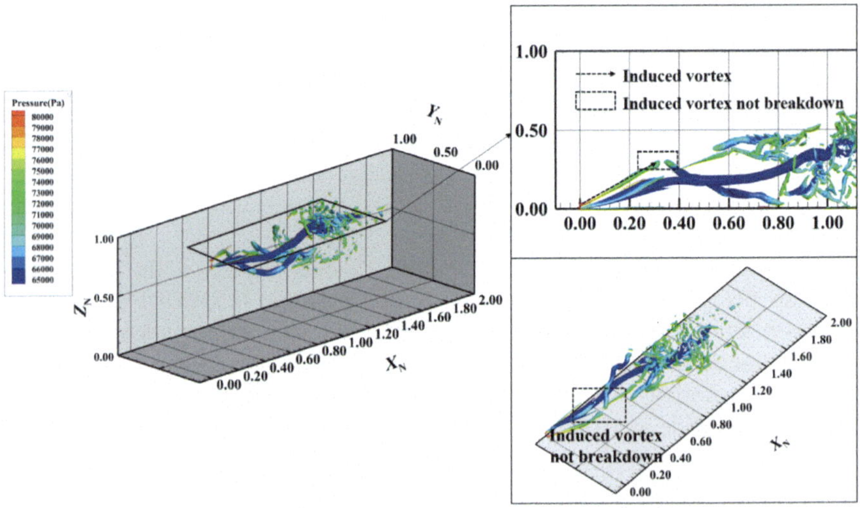

(b) DDES

Fig. 6.22 The transient vortex structures of the LES, DDES, and EDDES method

Further analysis of the flow field characteristics is conducted, as depicted in Fig. 6.23, which presents entropy contour plots along different streamwise sections for LES, DDES, and EDDES. From Fig. 6.23, it is evident that at the section $X_N = 0.3$, due to the action of K-H instability, the STLV has already formed and is entrained into the TLV. In the DDES results, the initial shear location exhibits characteristics of a shear flow across all streamwise sections without the resolved K-H

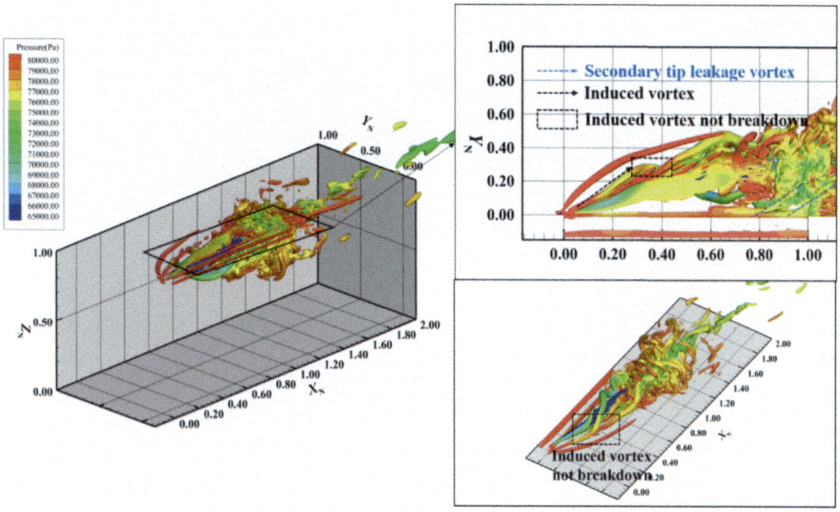

(c) EDDES

Fig. 6.22 (continued)

instability. Figure 6.23c illustrates that in the EDDES results, before $X_N < 0.7$, the flow field structures captured are similar to those of DES, with the STLV not being resolved. Consequently, in the absence of fluctuation energy injection, the IV does not breakdown and is entrained to the inner side of the TLV (in the Y-direction). At the section $X_N = 0.7$, the shear flow begins to be gradually perturbed, and the K-H instability is unlocked. The STLV starts to form near the section at $X_N = 0.8$, as indicated by the black dashed arrow. Subsequently, driven by the fluctuations of the STLV, as shown in the section $X_N = 0.88$ of Fig. 6.23c, the core of the IV begins to expand and breakdown, which is indicated within the red circle.

Figure 6.24 illustrates the turbulent kinetic energy, $k = \frac{1}{2}(\overline{u'u'} + \overline{v'v'} + \overline{w'w'})$, at various streamwise sections captured by the LES, DDES, and EDDES methods. It can be observed that, in comparison to the LES method, both the DDES and EDDES methods predict a lower level of turbulent kinetic energy, particularly within the range of $X_N < 0.7$. This implies that these two methods capture insufficient turbulence fluctuations upstream of the leakage flow, which are not adequate to induce shear layer instability and thus fail to resolve the fluctuations of the secondary leakage flow. Within the range of $X_N > 0.7$, the EDDES method resolves greater turbulent kinetic energy at the initial shear location than the DDES method, hence capturing the secondary leakage vortex within the range of $X_N = 0.7$–0.8.

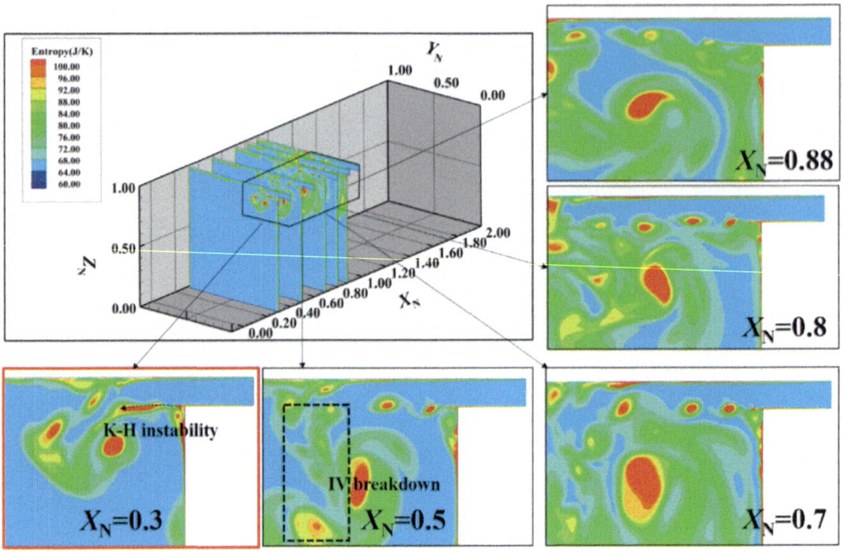

(a) LES

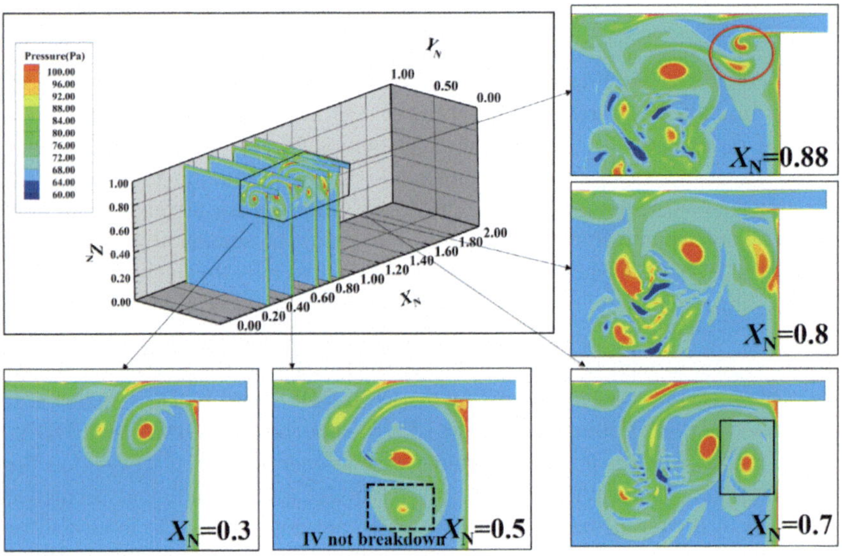

(b) DDES

Fig. 6.23 The entropy contour of LES, DDES, and EDDES, as well as the resolution of K-H instability

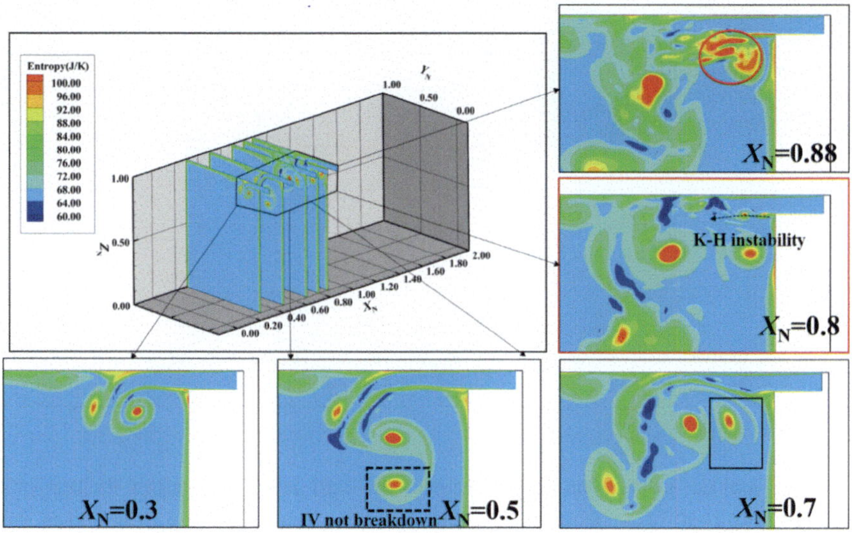

(c) EDDES

Fig. 6.23 (continued)

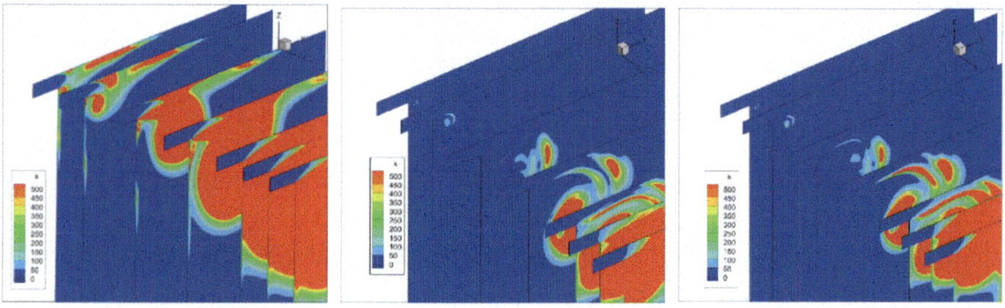

(a) LES (b) DDES (c) EDDES

Fig. 6.24 Prediction of turbulent kinetic energy by LES, DDES, and EDDES methods

According to the turbulence barycentric map [15], the analysis of turbulent anisotropy at the initial shear location of different streamwise sections by LES, DDES, and EDDES methods is compared. The selected measurement points are grid nodes in $X_{\mathrm{N}} = 0.34 \sim 0.85$ at the initial shear locations of five streamwise sections ($0 < Y_{\mathrm{N}} < 0.08$, $Z_{\mathrm{N}} = 0.95$), as shown in Fig. 6.25. From Fig. 6.26, it can be seen that the anisotropic characteristics of turbulence captured by the LES method are consistent at various streamwise locations, they are all transitioning from the initial leakage of dual-component turbulence to a single-component dominant turbulence, and finally trans into a composite turbulence state. This indicates that at the initial state of leakage in any cross-section, when the leakage flow just eject out of the gap, fluctuations in both directions are dominant; as the leakage flow leaves the gap, the

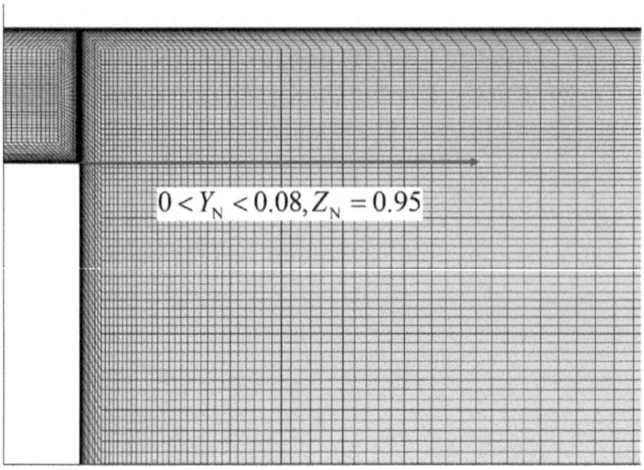

Fig. 6.25 Schematic diagram of data measurement points for turbulence barycentric maps in LES, DDES, and EDDES

dissipation effect of the boundary layer near the wall at $Y_N = 0$ causes fluctuations in a single direction to become dominant. Finally, as the leakage flow leaves the boundary layer and develops along the tangential direction, the LES results show a composite state of turbulence components. Comparing the results of DES, EDDES, and LES, at the initial streamwise cross-section, where $X_N < 0.5$, both of them transition from dual-component dominance to single-component dominance. At the cross-section of $X_N = 0.67$, EDDES gradually shows a trend towards developing a composite turbulence state. At the $X_N = 0.75$ cross-section, the results of EDDES are gradually approaching to the LES, indicating that the EDDES's analysis of turbulent anisotropy is close to the level of LES at this point, while the DES results still maintain single-component turbulence dominance. This corresponds to the EDDES capturing the microstructure of the STLV at this streamwise location as mentioned before. At the $X_N = 0.85$ cross-section, EDDES and DES are close to the results of LES near the location of $Y_N = 0.08$, but due to the breaking effect of the leakage vortex (see Fig. 6.12b, c), dual-component turbulence fluctuations become dominant, showing a significant difference from the LES results.

In summary, to achieve accurate predictions of DES-type methods in the flow in the tip region, it is necessary to ensure a thorough resolution of turbulent kinetic energy at the initial stage where the mainstream shears with the leakage flow upstream of the leakage. Although the EDDES method reduces the initial shear position's eddy viscosity by decreasing the subgrid scale, it still fails to achieve a resolution of turbulent kinetic energy upstream of the leakage. Compared with the LES method, the generation of secondary leakage vortices is delayed, and the IV are delayed in breaking. Compared with the DDES method, the EDDES method resolves STLV downstream of the leakage, which leads to a more intense breakdown of the IV, resulting in greater discrepancies in the predicted losses and TLV's time-averaged trajectories compared to the LES method.

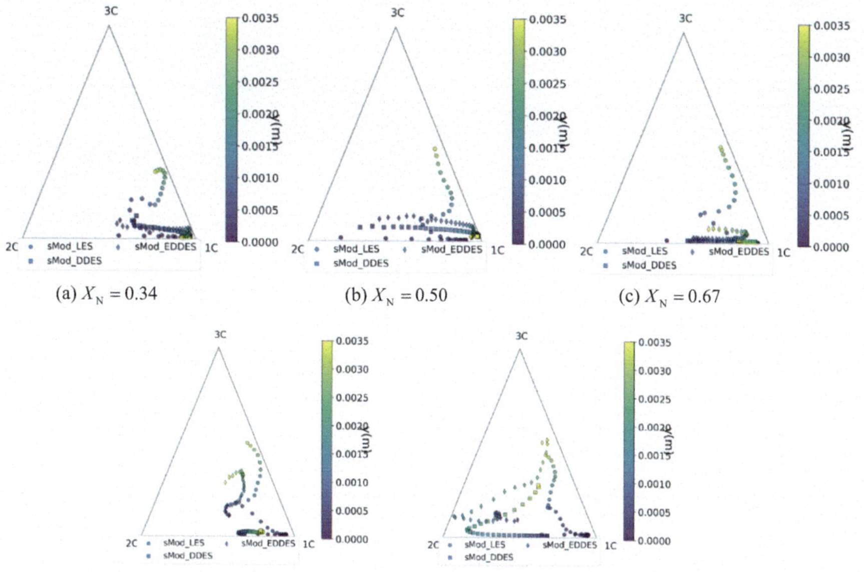

(a) $X_N = 0.34$ (b) $X_N = 0.50$ (c) $X_N = 0.67$

(d) $X_N = 0.75$ (e) $X_N = 0.85$

Fig. 6.26 Turbulence anisotropy barycentric map of the initial shear region for LES, DDES, and EDDES methods

6.4.4 Application in the Compressor Cantilever Stator Cascade

Computational Settings and Grid

In this study, we attempt to use the EDDES method to make predictions in the NPU01 cascade. The open-source SU2 code, is employed to solve the N-S equations. The upwind Roe scheme with MUSCL reconstruction is utilized to calculate the convective fluxes for to achieve second-order in space. To regulate dissipation within the LES region of the flow, an adapted variant of the Roe scheme is incorporated into SU2 [16]. Besides, an upwind scheme is applied to discretize the convective fluxes in the S-A equation. Regarding the time advancement, a dual-time stepping strategy with second-order implicit time integration was employed, and the physical time was set to 1×10^{-6} s, which keeps consistent with the settings in Sect. 6.2, based on the suggestions in Ref. [17].

Regarding to the boundary conditions, the inlet is specified with a uniform total temperature, and the total pressure is given with a radial distribution to achieve a boundary layer thickness of 10 mm in the incoming flow, consistent with the experiment. The inlet turbulence intensity is set as 1.5%, which is consistent with the experiment, while the inlet turbulent to laminar viscosity ratio 10%. The outlet is

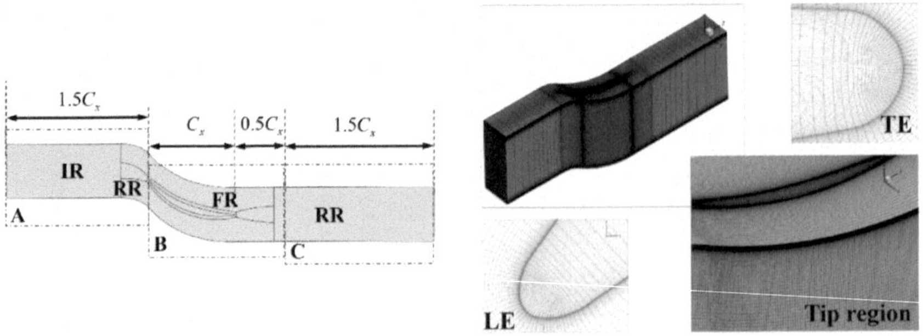

(a) The flow region in blade to blade view (b) The grid of the NPU01 in 3D view

Fig. 6.27 The diagram of the flow region and grid

Table 6.4 The grids in different regions

Region	A	B	C
Streamwise × spanwise × tangential	49 × 289 × 73	H: 217 × 289 × 49 O: 353 × 289 × 41	25 × 289 × 73

set with an average static pressure. The incoming Mach numbers $Ma_1 = 0.5$, leading the Reynolds number is 7.7×10^5 based on the chord length.

Structured O4H topology is employed for the main blade passage, and butterfly mesh topology is applied for modeling the blade tip gap. In the gap, there are 4,925,353 layers in the spanwise, tangential, streamwise directions, respectively. The wall distance of the first grid layer is 2×10^{-6} m, ensuring that $y^+ < 1$. According to the grid partition strategy, x^+ is less than 110 and z^+ is less than 120 in the focus region (FR region), i.e., the tip leakage and the corner separation area, which meets the requirements of the LES. The grid number is decreased to reduce computational in the RR region, as is shown in Fig. 6.27a. To further enhance computational efficiency, while ensuring that the grid number in the blade tip region is consistent with that in Sect. 6.2, in the region A and C, fewer grids are set, with particular emphasis on the region B of the blade passage and extending 0.5 times the axial chord length (C) from the outlet, resulting the total grid numbers as 13.81 million and the grid distribution in different regions is shown in Table 6.4.

The Accuracy of the EDDES Method in Tip Leakage Flow

To validate the accuracy of the EDDES method, Fig. 6.28 provides a comparison between the averaged streamline of the EDDES with the experimental oil flow pattern at $Ma_1 = 0.5, i = 0°$. Figure 6.28a shows the comparison of the suction surface at the blade tip region. The EDDES and experiments results exhibit good agreement at the LE separation and tip region of the suction surface. At the LE region, the yellow dashed line represents the LE separation line and it could be seen that the length of the leading-edge separation bubble is quite similar. At the tip region, the orange, red and black dashed line represent the position of the IV generation, the convergence line and the separation line caused by the scraping of the TLV and the suction surface, respectively. At the TE region, the green dashed circle represents the recirculation region. It could be seen that they all correspond well in both spanwise and streamwise ranges, indicating that the EDDES method provides accurate predictions for the span range of the tip leakage vortex.

Figure 6.28b shows the comparison of the tip endwall. The separation lines on the pressure side of both align well, while on the suction side, the convergence lines in EDDES slightly deviate toward the passage. However, the overall trend is consistent, indicating that the EDDES method slightly overpredicts the tangential extent of the tip leakage vortex. Furthermore, the positions where the leakage vortex, passage vortex, and induced vortex interact with each other (indicated by the black circles in Fig. 6.28b, manifested as deflections in the streamlines and accumulation in the oil flow coatings) also correspond well.

Overall, based on the comprehensive comparison above, the EDDES method demonstrates good accuracy in predicting the tip leakage flow.

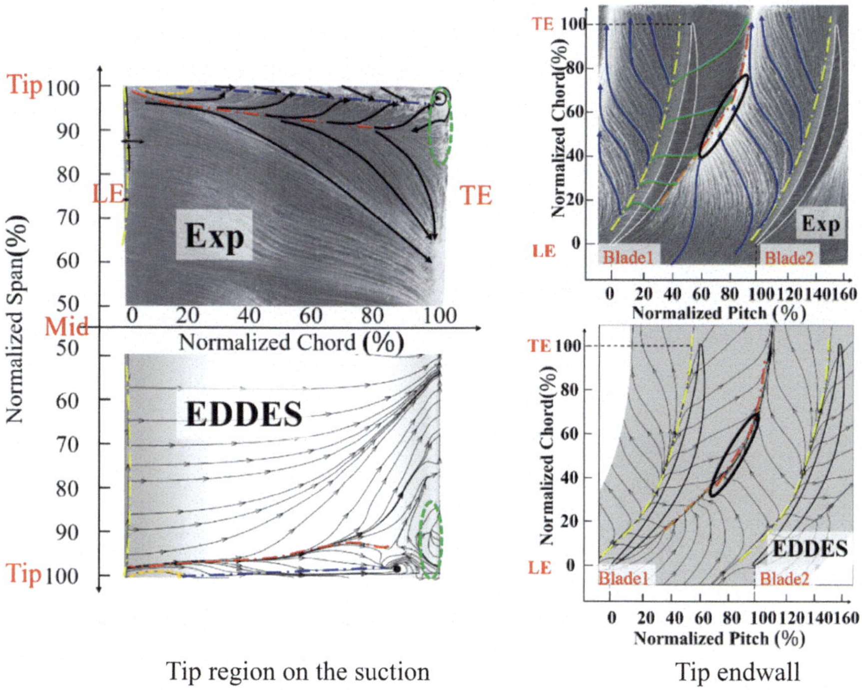

Fig. 6.28 Comparison between the averaged streamline of the EDDES with the experimental oil flow pattern at $Ma_1 = 0.5, i = 0°$

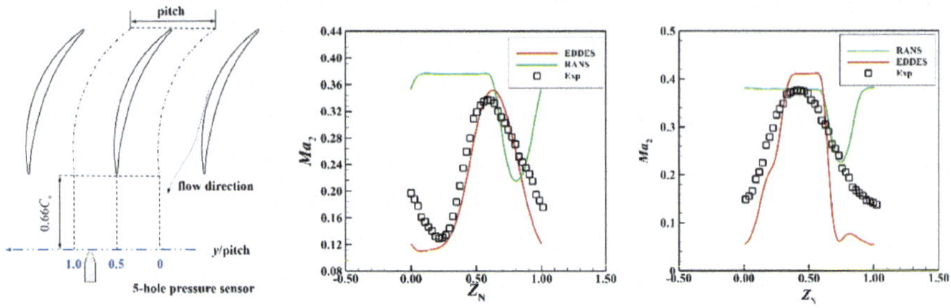

(a) The schematic diagram of the measurement position (b) $i = 0°$ (c) $i = 4°$

Fig. 6.29 Comparison of the Ma_2 distribution between the experiment, RANS and EDDES results

Figure 6.29b, c present the comparison of the distributions of the outlet Mach number, i.e., Ma_2 at $i = 0°$ and $4°$, respectively. As shown in Fig. 6.29a, the five-hole probe is located at the 166%C plane in the streamwise and at 85% span of the blade in the spanwise. It could be seen that the predictive accuracy of EDDES is higher than the RANS results in the tip region.

Flow Field Structure and Sources of Losses in Tip Region

1. The Vortex Structure of the Tip Region

Figure 6.30 depicts the transient vortex structure in the flow field (iso-surface $Liutex_{mag} = 8000$ contoured by Mach number). It could be observed in Fig. 6.30a that the entire flow field along the span exhibits three relatively independent vortex systems: the tip vortex system located above approximately 65% of the span, the separation in the mid-span, and the blade root corner separation below 40% range of the span.

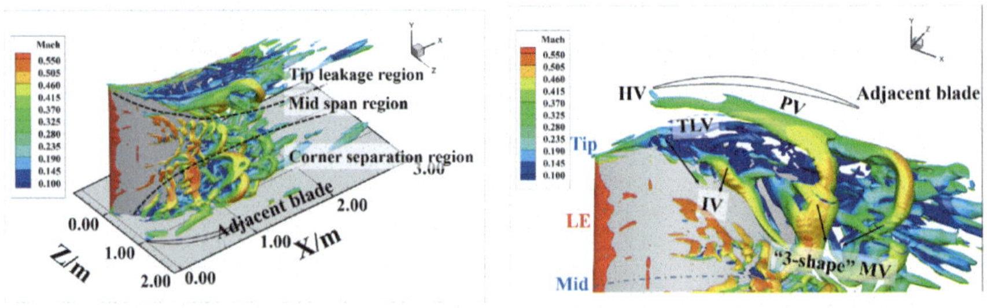

(a) The vortex structure of the entire flow filed (b) The vortex structure in the tip region

Fig. 6.30 The transient vortex structure in the flow field (iso-surface $Liutex_{mag} = 8000$ contoured by Mach number)

For a detailed examination, Fig. 6.30b provides a zoomed-in view of the blade tip area. It could be seen, the resolution of the vortex structures in the tip regions by EDDES method is significantly improved compared to that of the DDES method (as seen in Fig. 6.6b of Sect. 6.2). Furthermore, it could be seen that there are four distinct vortex structures:

(1) Tip Leakage Vortex (TLV): Originating at the $3\%C$, it is initially located close to the suction surface before approximately $25\%C$. Subsequently, the trajectory of the tip leakage vortex turns into passage, causing a reduction in fluid velocity around the vortex core. Low-energy fluid begins to appear on the inner side of the TLV near the suction surface tip. As the flow develops downstream, the low-energy fluid extends along both the spanwise and tangential ranges.
(2) Induced Vortex (IV): At the $25\%C$ axial chord length, when the vortex core of the TLV begin to detach from the suction surface and deflect into the passage, the airflow on the inner side near the suction surface induces shear on the low-energy fluid near the wall. This induces the generation of the IV. The IV develops along the suction surface, and as the previous branch of IV is about to detach from the suction surface, a new branch begins to generate.
(3) The pressure surface branch of the horseshoe vortex (HV) near the leading edge of adjacent blades gradually merges with the passage vortex (PV) as it develops downstream.
(4) "3-Shape" Mixed Vortex ("3-Shape" MV): The IV, as it develops downstream, gets stretched and extends into the space. One leg of the IV, directed upwards, interacts with the low-energy fluid near the blade surface, while the other leg, pointing backward, merges with the PV. This interaction results in the generation of the "3-shape" mixed vortex ("3-shape" MV), so named due to its shape resembling the Arabic numeral 3, as depicted in the figure, covering the spanwise range from 65 to 100%. Eventually, it completely detaches from the blade suction surface at the trailing edge and mixes with the wake.

2. The Loss of the Tip Region

Figure 6.31 shows the distribution of entropy generation rate and total pressure loss in the tip region along the flow direction. Seven quasi-S3 sections are extracted from the leading edge to the trailing edge to analyze the distribution of high-loss structures (areas with high entropy generation rate) and cumulative losses (areas with large total pressure loss) at different streamwise positions. Through Fig. 6.31a, it can be observed that before $20\%C$, the predominant contributor to losses in the tip region arises from the presence of the TLV. After that, the TLV deflects into the passage, forming IV and low-energy fluid near the suction surface. In the range from $20\%C$ to $78\%C$, the IV gradually separates from the suction surface and subsequently converges with the low-energy fluid (indicated by the black circular dash line in Fig. 6.30) and the PV, corresponding to regions of high loss in the flow field. Subsequently, it forms the "3-shape" MV and develops downstream. The high-loss region extends across the entire passage, reaching into the mid-blade region up to

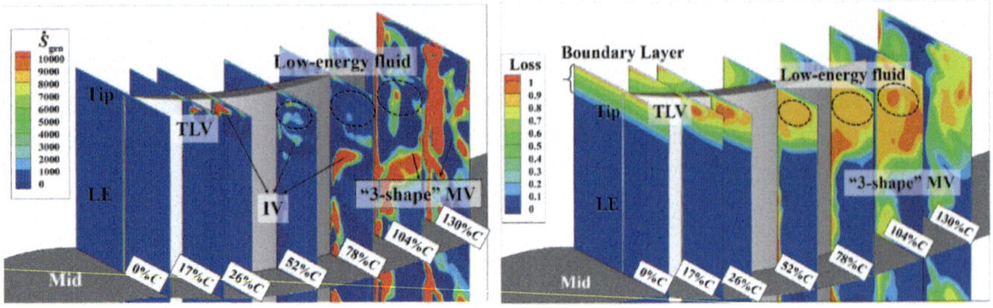

(a) Entropy generation rate (b) Total loss

Fig. 6.31 Entropy generation rate distribution and total pressure loss contour in the tip region at $Ma_1 = 0.5, i = 4°$

65% of the blade span, eventually detaching from the trailing edge and blending with the wake.

The total pressure loss shown in Fig. 6.31b represents the energy loss of the local position relative to the incoming airflow. Within the tip endwall boundary layer, the airflow experiences a reduction in velocity and a consequent decrease in energy due to viscous effects. The velocity of the fluid around the core of the TLV remains relatively low. Consequently, over the entire downstream range, the total pressure loss within the region covered by the TLV and its inner low-energy fluid is consistently high. Although the low-energy fluid has low velocity and insufficient energy, the level of momentum mixing induced by the fluctuations is limited, resulting in a relatively low loss-producing capability. Similarly, in the region where the "3-shape" MV begins to generate, the total pressure loss is not high. This is because the fluid velocity in this region is relatively high, and thus, there is no excessive loss of total pressure. In fact, the intense mixing between different vortex structures is the reason for the high-loss sources.

Therefore, it can be seen that, compared to total pressure loss, entropy generation rate has more advantages in analyzing the source of the high loss structures.

References

1. Wisler D C. Loss reduction in axial-flow compressors through low-speed model testing. Journal of Engineering for Gas Turbines and Power, 1985, 107(2): 354–363.
2. Yamada K, Kikuta H, Furukawa M, et al. Effects of tip clearance on the stall inception process in an axial compressor rotor//Proceedings of the ASME Turbo Expo 2013: Turbine Technical Conference and Exposition. San Antonio, Texas, USA: ASME, 2013: V06CT42A035.
3. Fukano T., Jang C. M. Tip clearance noise of axial flow fans operating at design and off-design condition. Journal of Sound and Vibration, 2004, 275(3–5): 1027–1050.
4. Smith J. T., House M. E. Internally generated noise from gas turbine engines: Measurement and prediction. Journal of Engineering for Power, 1967, 89(2): 117–185.
5. Su X., Ren X., Li X., et al. Unsteadiness of tip leakage flow in the detached-eddy simulation on a transonic rotor with vortex breakdown phenomenon. Energies, 2019, 12(4): 954.

6. Wu J. Improved delayed detached eddy simulation of compressor cascade tip leakage flow. International Journal of Aeronautical and Space Sciences, 2020, 22(1): 17–32.
7. Riera W., Marty J., Castillon L., et al. Zonal detached-eddy simulation applied to the tip-clearance flow in an axial compressor. AIAA Journal, 2016, 54(8): 2377–2391.
8. Spalart P. R. Young-person's guide to detached-eddy simulation grids. Contractor Report 211032, NASA, 2001.
9. Lilly D K. A proposed modification of the Germano subgrid-scale closure method. Physics of Fluids A: Fluid Dynamics, 1992, 4: 633–635.
10. Menter F. R., Kuntz M. Adaptation of eddy-viscosity turbulence models to unsteady separated flow behind vehicles//The Aerodynamics of Heavy Vehicles: Trucks, Buses, and Trains. Berlin, 2004: 339–352.
11. Celik I. B., Cehreli Z. N., Yavuz I. Index of resolution quality for large eddy simulations. Journal of Fluids Engineering, 2005, 127(5): 949–958.
12. Kay S., Marple S. L. Spectrum analysis—A modern perspective. Proceedings of the IEEE, 1981, 69(11): 1380–1419.
13. Chu B. T. On the energy transfer to small disturbances in fluid flow (Part I). Acta Mechanica, 1965, 1(1): 215–234.
14. Thomas D. E., Francisco P., Sean R. C., et al. SU2: An open-source suite for multiphysics simulation and design. AIAA Journal, 2016, 54(3): 828–846.
15. Banerjee S. S., Krahl R., Durst F., et al. Presentation of anisotropy properties of turbulence, invariants versus eigenvalue approaches. Journal of Turbulence, 2007, 8(1): 32.
16. Molina E. Hybrid RANS/LES calculations in SU2//23rd AIAA Computational Fluid Dynamics Conference. 2017.
17. Sun J., Liu Y., Lu L., et al. Control of corner separation to enhance stability in a linear compressor cascade by boundary layer suction. Procedia Engineering, 2014, 80: 380–391.

Chapter 7
Application of DES-Series Method in Inlet Distortion Generator Flow Field Analysis

7.1 Introduction

As military aircraft increasingly require enhanced capabilities for unconventional and supersonic maneuvers, the non-uniformity of engine intake parameter distribution becomes more prominent, such as the ingestion of ground vortices during takeoff [1], the separation of the inlet lip during high-angle-of-attack maneuvers and disturbances from shock waves in the inlet boundary layer during supersonic flight [2], or the increased separation flows around protuberant aircraft components. Furthermore, in consideration of aerodynamics, stealth, and structural requirements for the next-generation military aircraft, the development of various novel inlet designs, such as S-shaped [3] and semi-buried inlets [4], exacerbates the disruption of inlet airflow uniformity. Extensive research has indicated that inlet distortions significantly impact the aerodynamic performance of downstream compressors [5], particularly exerting a primary or even decisive influence on the engine aerodynamic stability [6], leading to a critical threat to the flight safety. Therefore, considering the influence of distortion is critical while evaluating engine aerodynamic stability. For the lack of comprehensive theoretical analysis methods, the current assessments of engine aerodynamic stability mainly rely on experiment [7].

The distortion simulator serves as an experimental apparatus deployed during ground testing of aircraft assemblies and components, facilitating the generation of a distorted flow field at the aerodynamic interface plane (AIP) within the compressor inlet. This setup aims to replicate the inlet distortion phenomena. Nowadays, extensive experimental and numerical efforts have been devoted to distortion simulators [8–14], including the mechanics underlying the formation of distorted flow fields, the impact of incoming flow parameters on the parameters of the AIP within distortion simulators, the characteristics of distortion flow fields and so on. Considerable strides have been made in understanding the steady-state distortion performance of distortion simulators. However, because of the intake distortion of the next-generation military fighter aircraft representing the dynamic characteristic of wide-ranging and

R. Li et al., *Detached-Eddy Simulation Series Methods for Compressor Internal Flow*,
https://doi.org/10.1007/978-981-96-0255-1_7

intense fluctuations, the dynamic distortion characteristic becomes one of the key parameters which have a significant impact on the aero-engine performance. Consequently, the way to use the distortion simulator to generate intake distortion mapping and dynamic distortion coefficient, which are consistent with the actual flight environment, is the key for total pressure experiment of the aero-engine and compressor on the ground testing bench.

A fundamental prerequisite for distortion testing lies in elucidating the regulatory relationship of the distortion simulator for the dynamic distortion characteristics at the AIP. Yet, dynamic distortion testing poses challenges, such as the exorbitant cost of total pressure probes, constrained sensor placement, prolonged experimental timelines, limited data acquisition, safety concerns and so on. For these challenges, numerical simulation emerges as the preferred way for forecasting the flow dynamics within distortion simulators. Notably, conventional URANS methods [15] can't predict the flow fields of the distortion simulator because the characteristic of the flow field is large-scale and these methods can't predict it accurately in principle which are inherently limited by their assumption of turbulence homogeneity. In contrast, DES-type methodologies, striking a balance between computational efficiency and precision, offer a promising way for investigating dynamic distortion within distortion simulators. Considering the superior grid-independence characteristic of DDES, this chapter adopts a DDES numerical approach.

This chapter endeavors to develop a DDES numerical simulation method applicable to dynamic flow fields within baffle-type distortion simulators. It examines classical chord-wise baffle configurations, axial series-connected composite baffles and circumferential parallel-connected composite baffles, in order to study the influence of distinct configurations and baffle depths on steady-state and dynamic distortion coefficients and their ratio. Additionally, the process of evaluating the three configurations regulatory capacity of the distortion coefficients offers a theoretical to improve the accuracy and rationality of the distortion tests.

7.2 Definition of Distortion Coefficient

Total pressure distortion primarily stems from disturbances in the airflow upstream and within the inlet duct during actual flight. Total pressure distortion exhibits complexity and variability, with its value at any given moment being a function $p^*(r, \theta, t)$ containing spatial and temporal variables. To quantitatively assess the extent of total pressure distortion in the design, development, and testing of aircraft engines, total pressure distortion coefficients are commonly employed to establish a link between the degree of total pressure distortion and the stability and performance of aircraft engines. However, due to variations in research methods, experimental setups, and approaches employed by different countries in investigating the impact of inlet total pressure distortion, a unified standard for total pressure distortion coefficients is lack. This chapter adopts the Russian standard for inlet total pressure distortion coefficients, in accordance with the "GJB Z 64A-2004 Guidelines for Assessment of Total Pressure Distortion at the Inlet of Aircraft Turbojet and Turbofan Engines".

7.2.1 Steady-State Distortion Coefficient

The steady-state distortion coefficient $\Delta\overline{\sigma}_0$ is a dimensionless parameter employed to quantify the relative disparity between the average total pressure recovery coefficient σ_{av} at the AIP and the average total pressure recovery coefficient σ_0 within the circumferential range of the low-pressure region θ^-.

$$\Delta\overline{\sigma}_0 = 1 - \frac{\sigma_0}{\sigma_{av}} \tag{7.1}$$

$$\sigma_0 = \frac{1}{\theta^-}\int_{\theta_1}^{\theta_2} \sigma_{r,av}(\theta)\mathrm{d}\theta \tag{7.2}$$

$$\sigma_{av} = \frac{1}{360}\int_{0}^{360} \sigma_{r,av}(\theta)\mathrm{d}\theta \tag{7.3}$$

where the $\sigma_{av}(\theta)$ is the radial average total pressure recovery coefficient at a specific azimuthal position at the AIP, namely

$$\sigma_{r,av}(\theta) = \frac{\int_{\overline{r}_{hub}}^{1} \sigma(\overline{r},\theta)2\overline{r}\mathrm{d}\overline{r}}{1-\overline{r}_{hub}^2} \tag{7.4}$$

Refer to the parameter Table 7.1 for the detailed definitions of the parameter.

The azimuthal range of the low-pressure zone θ^- is utilized to describe the angular parameter of the radial total pressure recovery coefficient $\sigma_{r,av}(\theta)$ relative to the surface-averaged total pressure recovery coefficient σ_{av} on the AIP. The typical distribution chart of the $\sigma_{r,av}(\theta)$ is depicted in Fig. 7.1, and the azimuthal range of the low-pressure zone can be defined by the starting angle θ_1 and ending angle of the low-pressure zone θ_2, namely

$$\theta^- = \theta_2 - \theta_1 \tag{7.5}$$

When multiple low-pressure zones are present on the AIP, each low-pressure zone necessitates the calculation of its respective azimuthal range θ_i^- and steady-state distortion coefficient $\Delta\overline{\sigma}_0$. θ^- on the AIP is derived by the sum of the θ_i^-. According to the value of the θ_i^-, the steady-state distortion coefficient of the AIP interface is determined as follows:

Table 7.1 Definition parameters table for distortion coefficients

Notation	Definition
r	Radius of a measurement point on the AIP
r_{hub}	Hub radius
r_{tip}	Tip radius
$\overline{r}$	Dimensionless radius, namely $\overline{r} = r/r_{tip}$
$\overline{r}_{hub}$	Hub ratio, namely $\overline{r}_{hub} = r_{hub}/r_{tip}$
θ	Polar angle on the AIP (°)
p_0^*	Total pressure of uniform inflow upstream of the inlet (Pa)
P^*	Total pressure (Pa)
$\overline{P^*}$	Time-averaged total pressure, namely $\overline{P^*} = \frac{1}{T}\int_0^T P^* \mathrm{d}t$ (Pa)
$(\overline{P^*})_{av}$	Surface time-averaged total pressure (Pa)
T	Sampling time, which is greater than the longest period of airflow fluctuations (s)
θ^-	Total circumferential range of low-pressure zones on the AIP
σ	Total pressure recovery coefficient, namely $\sigma = \overline{p^*}/p_0^*$
σ_0	Average total pressure recovery coefficient within θ^-
σ_{av}	Surface-averaged total pressure recovery coefficient
$\sigma_{r,av}(\theta)$	Radial averaged total pressure recovery coefficient at circumferential position θ
$\Delta\overline{\sigma}_0$	Steady-state distortion coefficient
ξ_{av}	Surface averaged dynamic distortion coefficient
$\xi_{r,av}(\theta)$	Radial averaged dynamic distortion coefficient at circumferential position θ
$(\Delta p^*)'_{RMS}$	Total pressure root mean square (Pa)

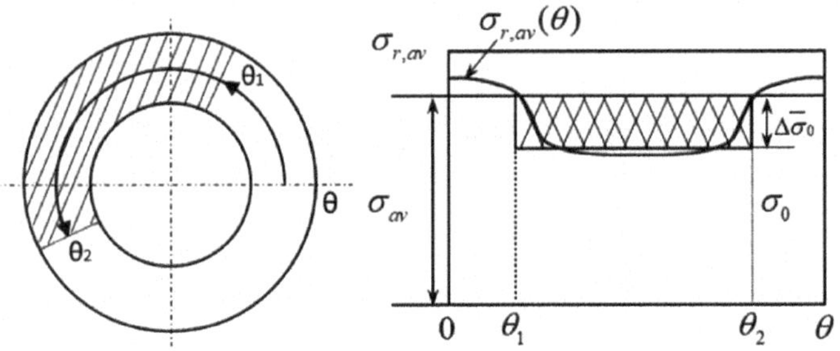

Fig. 7.1 The azimuthal distribution of the radial average total pressure recovery coefficient $\sigma_{r,av}(\theta)$

When $\theta_i^- > 60°$, the steady-state distortion coefficient of the distorted flow field is equal to the largest value of a specific parameter among all low-pressure zones.

When $\theta_i^- \leq 60°$, first compare the value of the $\left(\Delta\overline{\sigma}_0 \times \theta^-\right)/60°$ among the low-pressure zones, and select the maximum as the steady-state distortion coefficient of the distorted flow field.

When both of the aforementioned conditions are present simultaneously, apply the respective calculation method and select the maximum as the steady-state distortion coefficient of the distorted flow field.

7.2.2 Dynamic Distortion Coefficient

The dynamic distortion coefficient represents the temporal evolution of total pressure non-uniformity on the AIP. In engineering applications, the averaged dynamic distortion coefficient across the interface, ξ_{av}, serves as a quantitative representation of this total pressure pulsation characteristic. The process of calculating the ξ_{av} involves calculating the radial average of the $\xi(\overline{r}, \theta)$ at a fixed point $(\overline{r}, \theta)$ first, followed by computing the azimuthal average.

$$\xi_{av} = \frac{1}{360^\circ} \int_0^{360^\circ} \xi_{r,av}(\theta) \mathrm{d}\theta \tag{7.6}$$

$$\xi_{r,av}(\theta) = \frac{\int_{\overline{r}_{hub}}^1 \xi(\overline{r}, \theta) 2\overline{r} \mathrm{d}\overline{r}}{1 - \overline{r}_{hub}^2} \tag{7.7}$$

$$\xi(\overline{r}, \theta) = \frac{(\Delta p^*)'_{RMS}}{\left(\overline{p^*}\right)_{av}} \tag{7.8}$$

where, $(\Delta p^*)'_{RMS}$ is the total pressure root mean square, and $(\overline{p^*})_{av}$ is the time-averaged total pressure on AIP.

$$\left(\Delta p^*\right)'_{RMS} = \sqrt{\frac{1}{T} \int_0^T \left(p^* - \overline{p^*}\right)^2 \mathrm{d}t} \tag{7.9}$$

$$\left(\overline{p^*}\right)_{av} = \frac{1}{360^\circ \times \left(1 - \overline{r}_{hub}^2\right)} \int_{\overline{r}_{hub}}^{1} \int_{360^\circ}^{0^\circ} \overline{p^*}(\overline{r}, \theta) 2\overline{r} d\theta \mathrm{d}\overline{r} \tag{7.10}$$

In actual flight conditions, the complexity and variability of the incoming airflow to the engine are significantly influenced by both the increased maneuvering conditions of the aircraft and changes in the inlet structure. This leads to a range of variations in the ratio between the steady and dynamic distortion coefficients at the AIP (Aerodynamic Interface Plane) interface. For example, in 2020, Jiang et al. [16] installed an inlet distortion measurement system at the outlet section of the intake of a certain fighter aircraft during flight tests in China. The measurements were conducted under various typical flight conditions, including small maneuvers such as steady level flight and climbing, as well as high-maneuver, high-g maneuvers such as half-rolls and loops. The results indicated that the ratio of steady-state to dynamic distortion at the AIP interface ranged from 57.1 to 316%. To more accurately reflect actual flight conditions in a ground test environment, it is necessary for the dynamic distortion index to cover a sufficiently wide range of the overall distortion index in the total pressure distortion simulator. Therefore, this study focuses on the ratio between the dynamic distortion index and the steady-state distortion value.

7.3 Numerical Simulation of Flow Field in Inlet Distortion Generator Based on DDES Method

7.3.1 Physical Model

The baffle simulator is a distortion simulator recommended for use by Russia and is primarily composed of solid baffles of various shapes. When these solid baffles are inserted into the intake duct, they create a low-pressure region influenced by the baffle behind it, and a relatively unaffected high-pressure region. Simultaneously, low-energy fluid at the edges of the baffles mixes with the main flow, generating turbulent pressure fluctuations with randomly varying frequency and amplitude. This results in a flow field at a certain distance downstream from the baffle, characterized by steady-state pressure loss (steady total pressure distortion), dynamic pressure variations (dynamic total pressure distortion), and a distortion pattern.

This study investigates three types of baffle distortion simulators: the Single-Baffle Distortion Generator (SD generator), the Tandem Double-Baffle Distortion Generator (TDD generator), and the Circumferential Multi-Baffle Distortion Generator (CMD generator). Using the Delayed Detached Eddy Simulation (DDES) method, which offers higher fidelity in separated flow fields, and validating the accuracy of the numerical methods against unsteady experimental results, this research provides a detailed analysis of the dynamic distortion characteristics and their influencing factors. This section first presents the physical models of the three distortion simulators.

(1) **Single-Baffle Distortion Generator (SD Generator)**

The Single-Baffle Distortion Generator offers advantages such as easy installation, simple structure, good versatility, and a wide range of disturbance. Russia has several decades of experience in the engineering development and use of this type of simulator, which has been extensively developed and widely applied in various total pressure distortion testing devices. It is also recommended by the Chinese military standard GJB/Z 64A-2004 "Guidelines for the Assessment of Inlet Total Pressure Distortion in Aircraft Turboshaft and Turbofan Engines" as the primary testing device for simulating engine inlet total pressure distortion.

To validate the methodology, the dimensions of the SD generator used in this study are based on the actual test apparatus provided by a research institute, as shown in Fig. 7.2. The primary geometric parameters of the crescent-shaped baffle total pressure distortion simulator include the intake duct diameter D, baffle depth H, distance L between the baffle and the AIP, and baffle thickness δ. The baffle is inserted from above the duct, and its depth H is measured from the upper edge of the duct to the lower edge of the baffle. Notably, baffle depth serves as a crucial adjustable parameter in the experiment. This chapter will analyze three scenarios with $H = 0.2D, 0.3D, 0.4D$ as outlined in Table 7.2.

(2) **Tandem Double-Baffle Distortion Generator (TDD Generator)**

Building on the Single-Baffle Distortion Generator, a tandem double-baffle distortion generator was employed to explore ways to adjust the distortion coefficient ratio. This configuration forms the Axial Tandem Double-Baffle Distortion Generator, as shown in Fig. 7.3. Its geometric characteristics are primarily described by the relative depth of the baffles H/D, the relative circumferential position of the two baffles α, and the axial distance between them ΔL.

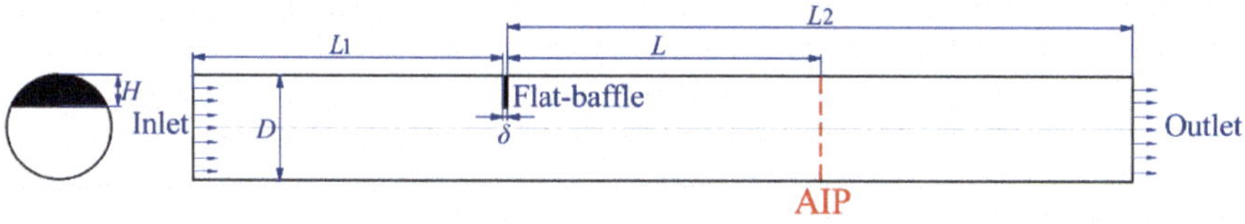

(a) the flat baffle in the cylinder (black) (b) stream-wise position of the flat baffle

Fig. 7.2 Configuration of single-baffle distortion generator

Table 7.2 The geometric parameters of the single-baffle distortion generator

Geometric parameters	Value	Geometric parameters	Value
Diameter D (m)	0.308	Baffle thickness δ (m)	0.010
The distance from the inlet to the baffle L_1 (m)	3D	AIP distance L (m)	3D
The distance form the baffle to the outlet L_2 (m)	6D	Baffle depth H (m)	0.2D, 0.3D, 0.4D

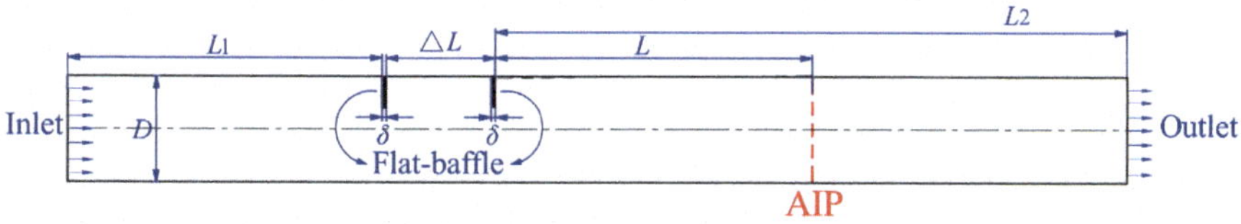

Fig. 7.3 Configuration of tandem double-baffle distortion generator

As a preliminary investigation, this study focuses on a numerical analysis of different baffle depths. Similar to the Single-Baffle setup, the study explores flow field distortion characteristics for baffle depths of $0.2D$, $0.3D$, $0.4D$, with axial spacings between the front and rear baffles of $1D$ and $0.5D$. The remaining parameters are detailed in Table 7.3. The front and rear baffles change synchronously, meaning that the depth H of both baffles remains the same throughout the experiments.

(3) **Circumferential Multi-baffle Distortion Generator (CMD Generator)**

Building on the principles of the Single-Baffle Distortion Generator, a circumferential multi-zone baffle distortion simulator was developed to study the characteristics of distortion flow fields where multiple small-angle low-pressure regions coexist. This setup also explores methods for adjusting the distortion coefficient ratio. The configuration consists of several small-sector baffles evenly distributed along the circumference, as shown in Fig. 7.4. The geometric features of the circumferential multi-zone combination baffle configuration are primarily described by the relative depth of the baffles H/D, the number of baffles N, and the sector angle of the baffles β. In this preliminary investigation, this chapter focuses solely on investigating the distortion characteristics at the AIP for baffle depths $0.2D$, $0.3D$, $0.4D$, with the baffle depths synchronously varied. The number of baffles is $N = 4$ and the sector angle $\beta = 20°$. Specific parameters are detailed in Table 7.4.

Table 7.3 The geometric parameters of the axial series combination baffle total pressure distortion simulator

Geometric parameters	Value	Geometric parameters	Value
Diameter D (m)	0.308	Baffle thickness δ (m)	0.010
Distance upstream of the baffle L_1 (m)	$3D$	AIP distance L (m)	$3D$
Distance downstream of the baffle L_2 (m)	$6D$	Baffle depth H (m)	$0.2D$, $0.3D$,$0.4D$
Baffle axial distance ΔL (m)	$1D$, $0.5D$	Baffle relative circumferential position α (°)	0

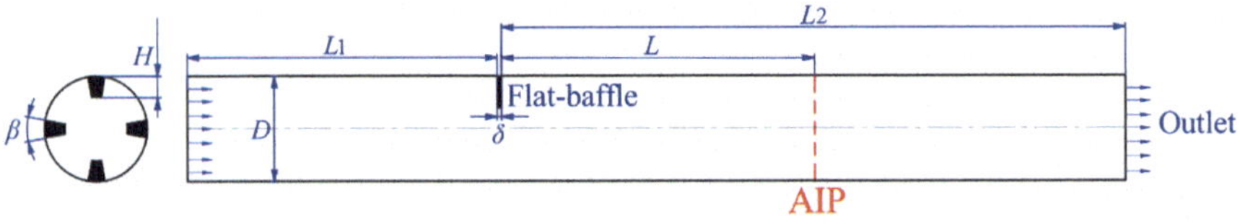

Fig. 7.4 Configuration of circumferential multi-baffle distortion generator

Table 7.4 The geometric parameters of the circumferential multi-zone combination baffle total pressure distortion simulator

Geometric parameters	Value	Geometric parameters	Value
Diameter D (m)	0.308	Baffle thickness δ (m)	0.010
Distance upstream of the baffle L_1 (m)	3D	AIP distance L (m)	3D
Distance downstream of the baffle L_2 (m)	6D	Baffle depth H (m)	0.2D, 0.3D, 0.4D
The number of baffles N	4	The sector angle of the baffles β (°)	20

7.3.2 Grid Generation

The mesh generation process must first ensure grid independence for RANS simulations, followed by localized refinement in critical research areas before conducting DDES numerical simulations. For the three aforementioned physical models, the geometric model of the baffle distortion simulator was created using Unigraphics software, and the mesh was generated using ICEM CFD software. To ensure mesh quality and orthogonality, a multi-block structured butterfly mesh was employed, as shown in Fig. 7.5. The mesh was independently refined near the solid walls, in front of and behind the baffle, and below the baffle. The height of the first layer of the mesh near the wall was set to 0.01 mm to maintain a wall $y+$ value around 1. The computational domain from the baffle to the AIP interface was identified as the key area of study.

Taking the typical Single-Baffle Distortion Generator as an example, a grid independence study was conducted based on RANS numerical simulations with a baffle depth of $H = 0.2D$ and an incoming flow Mach number of $Ma = 0.2$. RANS simulations were performed with grid sizes of 0.93, 1.38, and 1.96 million cells, with the difference in grid size being the density of the mesh between the baffle and the AIP interface. A third-order MUSCL scheme was used for spatial discretization, a second-order implicit scheme for temporal discretization, and the Spalart–Allmaras (S-A) turbulence model was employed. The simulation conditions included a baffle depth of $H = 0.2D$, an incoming flow Mach number of $Ma = 0.2$, and axial intake flow.

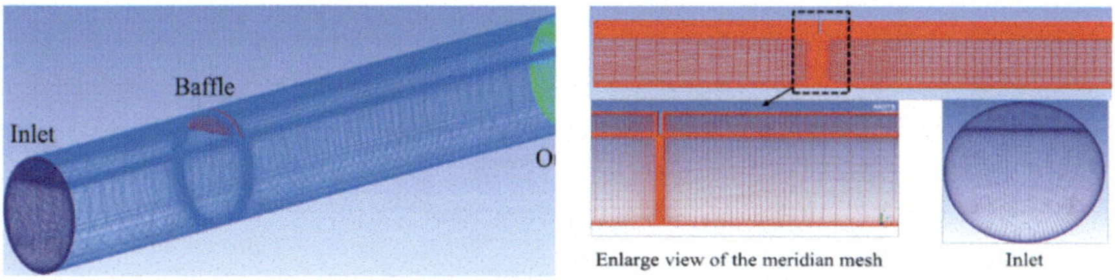

Fig. 7.5 Computational grid for RANS simulation of SD generator

The numerical simulation results, as depicted in Table 7.5 and Fig. 7.6, reveal that once the grid resolution surpasses 1,380,000 cells, key computed parameters including the surface-averaged total pressure recovery coefficient σ_{av}, the extent of low-pressure regions θ^{-}, and the steady-state distortion coefficient $\Delta\overline{\sigma}_0$ exhibit minimal variations. Thus, for the RANS computational grid, achieving a grid resolution of approximately 1,380,000 cells suffices to meet the criterion of grid independence.

The DDES numerical method imposes stricter requirements on grid resolution. Based on the analysis of RANS computational outcomes and the extensive engineering expertise within the research group, coupled with pertinent literature and *FLUENT* theoretical manuals, refining grids in regions exhibiting pronounced gradients in flow field physical parameters, particularly those distant from the wall behind the plate. Primary refinement regions of flow direction are between the baffle and the AIP. The circumferential and radial refinement regions are the areas of significant total pressure loss identified in the RANS results. Within these focal research areas, the following grid refinement principles are adhered to:

(1) Strive for uniform grid scales in all three dimensions.
(2) Maintain a grid scale transition ratio between sparse and dense regions below 2:1.
(3) Ensure that the maximum grid scale in all three dimensions doesn't exceed 0.1 times the thickness of the undisturbed boundary layer.

Table 7.5 Grid independence analysis of the insert plate total pressure distortion simulator

Grid resolution (million)	Surface-averaged total pressure recovery coefficient σ_{av}	Low-pressure region θ^{-} (°)	Steady-state distortion coefficient $\Delta\overline{\sigma}_0$ (%)
0.93	0.992	164.1	0.47
1.38	0.992	134.8	0.58
1.96	0.992	134.8	0.58

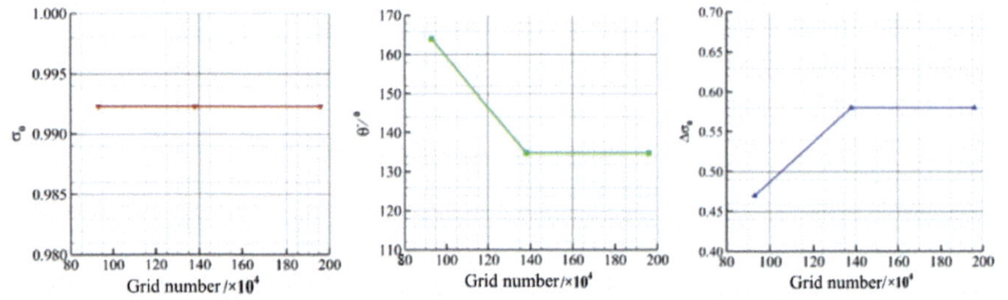

(a) Surface-averaged total pressure recovery coefficient σ_{av}. (b) Low-pressure region θ^{-}. (c) Steady-state distortion coefficient $\Delta\overline{\sigma}_0$.

Fig. 7.6 Grid independence assessment of the baffle total pressure distortion simulator

As depicted in Fig. 7.7a, the region extending from behind the baffle to the AIP is the focal research area, partitioned into three distinct zones. Zone 1 and Zone 2 correspond to regions characterized by elevated turbulent viscosity based on the RANS outcomes. Within these zones, the grid is configured to approximate cuboids and the maximum grid scale in all three dimensions doesn't exceed 0.05 and 0.1 times the thickness of the undisturbed boundary layer. Zone 3 features a grid scale twice that of Zone 1 and Zone 2 along the flow direction to facilitate a gradual transition from dense to sparse grids. To ensure grid quality and orthogonality, employ a multi-block structured hexahedral grid in the circumferential direction, as delineated in Fig. 7.7b, c. To mitigate computational overhead, taper the inlet and outlet extension sections in all three dimensions as illustrated in Fig. 7.7b, and refine the circumferential gird behind the baffle form Zone 1 to Zone 3 as illustrated in Fig. 7.7c.

Upon final determination to employ the DDES, the mesh configuration for the SD generator is illustrated in Fig. 7.8, comprising approximately 5 million in total. For the axial series combination baffle distortion simulator in Fig. 7.9, an extension from baffle 1 to baffle 2 was added to the crescent-shaped mesh, resulting in a total mesh count of around 7.83 million. Similarly, the circumferential parallel combination baffle distortion simulator mesh in Fig. 7.10, exhibits symmetric distribution along the circumferential axis. To mitigate computational complexity, a computational

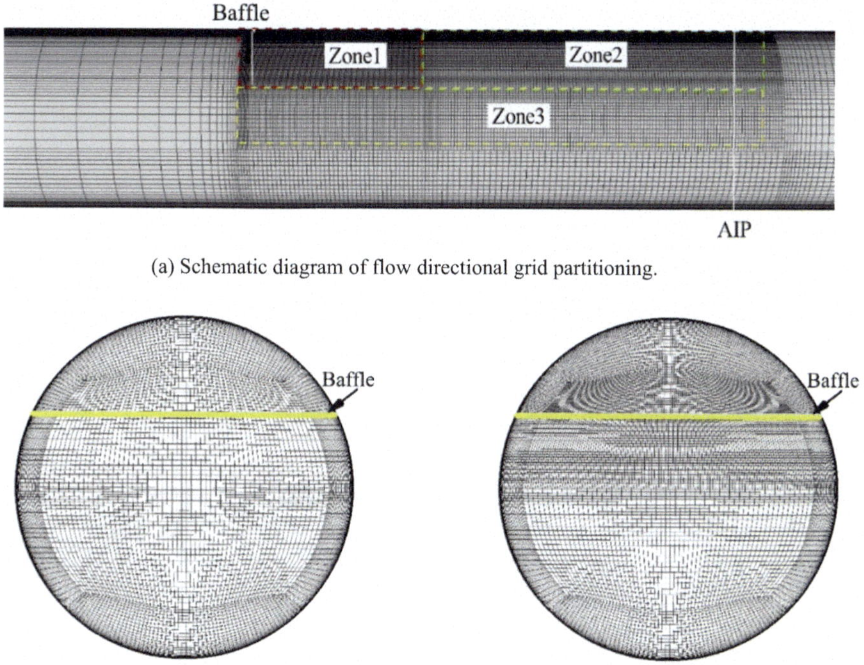

(a) Schematic diagram of flow directional grid partitioning.

(b) Circumferential grid in the inlet extension section. (c) Circumferential grid behind the baffle.

Fig. 7.7 Grid partitioning for DDES

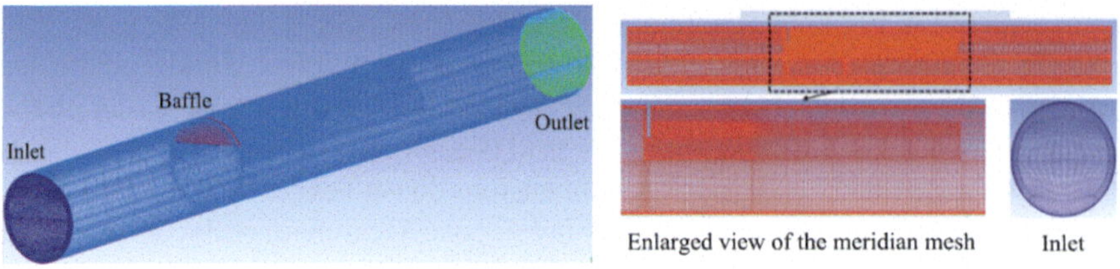

Fig. 7.8 Computational grid of SD generator for DDES

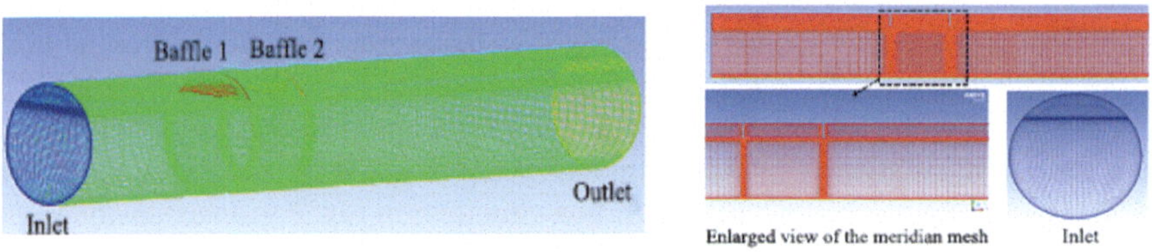

Fig. 7.9 Computational grid of TDD generator for DDES

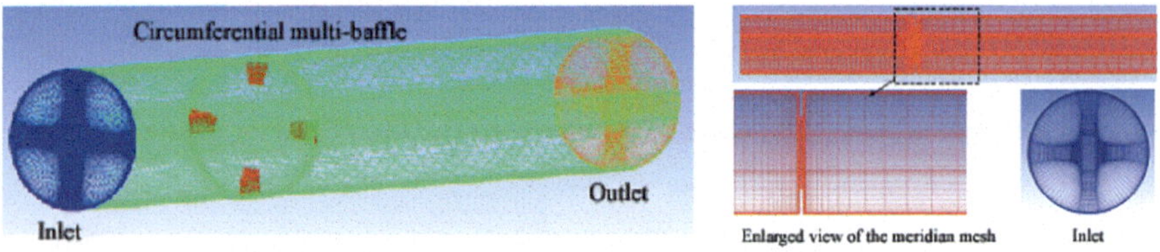

Fig. 7.10 Computational grid of CMM generator for DDES. Circumferential parallel combination baffle distortion simulator DDES computational mesh

domain encompassing a 180° range in the circumferential direction was selected, yielding a total mesh count of approximately 7.55 million.

7.3.3 *The Selection of Near-Wall Turbulence Models*

According to the introduction in Chap. 3, the DDES method is developed based on RANS turbulence models, with RANS models still being utilized in the near-wall region. Therefore, the selection of an appropriate turbulence model is paramount for ensuring the accuracy of the results. Large-scale separated flow exists downstream of the baffle in the baffle distortion simulator, and as of now, there is no universally accepted standard for turbulence models tailored to such flows. Hence, this subsection focuses on a crescent-shaped baffle distortion simulator with a baffle depth of $H = 0.2D$ and an incoming flow Mach number of $Ma = 0.2$. Three turbulence models are compared: the widely adopted one-equation S-A model in aerospace applications, the popular Shear-Stress Transport (SST) model among two-equation models, and the Realizable k–ε model known for its better prediction of flow separation. These models are employed to numerically compute the steady flow field, and the most suitable turbulence model is determined through comparison with experimental data.

Table 7.6 presents a comparative analysis between the steady-state computational outcomes of the three turbulence models and experimental data. For the surface-averaged total pressure recovery coefficient σ_{av}, all turbulence models demonstrate satisfactory agreement with experimental data, with a maximum error of 0.16%. For the low-pressure region parameter θ^-, all turbulence models compute angles lower than the experimental measurements, with the S-A model exhibiting the smallest deviation of 12%. For the steady-state distortion coefficient $\Delta\overline{\sigma}_0$, both the S-A model and Realizable k–ε model yield values below those obtained experimentally, while the SST model generates values exceeding experimental data. Among them, the S-A model displays the least deviation from experimental results, with an error of 6.4%. In conclusion, the S-A turbulence model is more suitable for the baffle distortion simulator. Therefore, employing the DDES numerical method based on the S-A model is recommended (Fig. 7.11).

Table 7.6 Crescent-shaped baffle: comparative analysis of numerical results from three turbulence models with experimental data

	Surface-averaged total pressure recovery coefficient σ_{av}	Low-pressure region θ^- (°)	Steady-state distortion coefficient $\Delta\overline{\sigma}_0$ (%)
Experiment	0.9910	153.3	0.62
S-A	0.9920	134.8	0.58
SST	0.9917	108.1	0.96
Realizable k–ε	0.9926	129.2	0.38

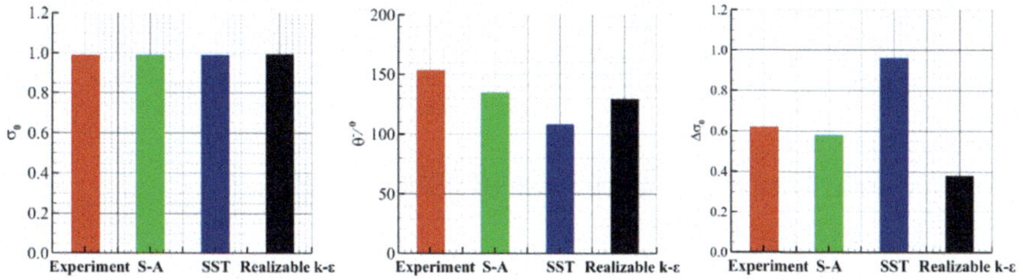

(a) Surface-averaged total pressure recovery coefficient σ_{av}
(b) Low-pressure region θ^-
(c) Steady-state distortion coefficient $\Delta\bar{\sigma}_0$

Fig. 7.11 SD generator: comparative analysis of numerical results from three turbulence models with experimental data

7.3.4 The Spatiotemporal Discretization Scheme and Boundary Condition

The spatiotemporal discretization scheme significantly influences the stability, accuracy, and computational resources consumption of numerical results. Spatial discretization employs a second-order central differencing scheme, while temporal discretization adopts a two-time-level method. The physical time step is $\Delta t = \overline{\Delta}/V_{max}$, where the V_{max} is typically set at least 1.5 times the inlet velocity, and the $\overline{\Delta}$ represents the average grid scale of the primary research domain. In this study, the physical time step is 5×10^{-5} s, with a CFL number of 1.0. The maximum iteration steps for the pseudo-time step are set to 65, and the total computational physical time is 0.5 s.

The boundary conditions of DDES are in accordance with those of RANS. Specifically, at the inlet, uniform total temperature and total pressure boundary conditions prevails. At the outlet, a uniform static pressure condition prevails. The physical walls are characterized by adiabatic and no-slip boundary conditions. To regulate the inlet Mach number, adjustments are made to the outlet back pressure. The initial field for DDES unsteady simulations is typically derived from the converged flow field resulting from the RANS simulation.

The convergence criteria for RANS simulations entail several conditions:

(1) The global residual decreases to a constant value or decreases gradually.
(2) Significant periodic fluctuations emerge, and the relative error between the inflow and outflow mass flow rates remains below 0.1%, with the mass flow rate stabilized.
(3) Other overall performance parameters such as inflow Mach number, mass flow rate, and main flow velocity converge to constant values with increasing iteration counts.

For DDES unsteady simulations, the convergence standard is as follows: the variation in time-averaged mean and variance of the unsteady total pressure at monitoring points, which are set on the AIP, remains within 1%.

7.4 Verification of the Accuracy of Methods Based on Experimental Data

The research institute has comprehensive distortion coefficient measurement data for a SD generator with a baffle depth of $H = 0.2D$, which serves as valuable experimental data for validating the accuracy of numerical results. During the experimental process, both steady-state and dynamic total pressures were acquired using probes and the positions as illustrated in Fig. 7.12. Due to constraints in measurement techniques, the obtained data exhibits relatively low spatial resolution. On the AIP, steady-state total pressure measurements were obtained using total pressure rakes positioned

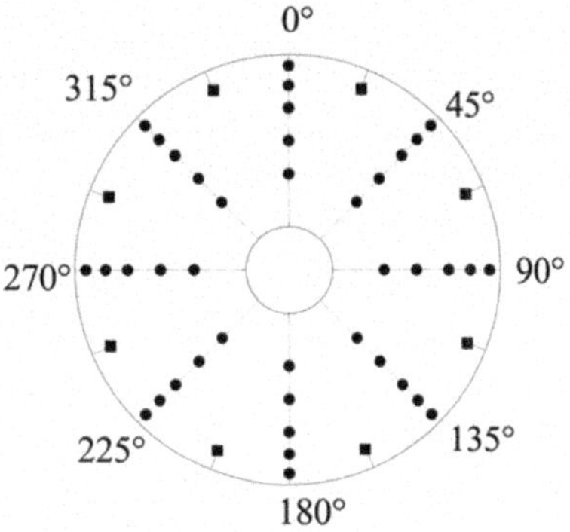

Fig. 7.12 The arrangement diagram of steady-state total pressure probes and dynamic pressure probes on the AIP

at 45° intervals circumferentially, depicted by circular points in Fig. 7.12. Radially, the measurement points were equally distributed, totaling 40 points. Dynamic total pressure measurements were sparser, with 8 measurement points placed radially at a position of $0.9D$, represented by square points in Fig. 7.12. To ensure numerical results' comparability with experiments, the distortion coefficient is computed at the same positions as the experimental measurement points.

7.4.1 Verification of Steady-State Numerical Accuracy

Conducted numerical simulations of steady-state flow fields for four inlet Mach numbers, $Ma = 0.2, 0.3, 0.4, 0.5$. A comparative analysis between numerical and experimental results is presented in Table 7.7 and Fig. 7.13. The error of surface-averaged total pressure recovery coefficient σ_{av} shows significant discrepancies only at $Ma = 0.5$, peaking at 1.7%. The error of the low-pressure region range θ^- diminishes with increasing Mach number, peaking at 12%. The error of steady-state distortion coefficient $\Delta\overline{\sigma}_0$ displays an increasing trend with higher Mach numbers, which is 82% at $Ma = 0.5$ while only 6.5% at $Ma = 0.2$. Overall, the discrepancies between experimental and numerical results enlarge with rising Mach numbers, attributed to heightened turbulence anisotropy in separated flow. Nonetheless, the general trend remains consistent. To ensure computational accuracy, subsequent research focuses on the $Ma = 0.2$ condition.

7.4.2 Verification of Dynamic Numerical Simulation Accuracy

The dynamic distortion coefficient of a SD generator is typically quantified by the mean dynamic distortion coefficient of total pressure pulsations ε_{av}. In this subsection, utilizing a baffle depth of $H = 0.2D$ and $Ma = 0.2$ condition as an

Table 7.7 Comparison table of steady-state distortion coefficient between numerical and experimental results for a 0.2*D* SD distortion simulator

Ma	Surface-averaged total pressure recovery coefficient σ_{av}		Low-pressure region θ^- (°)		Steady-state distortion coefficient $\Delta\overline{\sigma}_0$ (%)	
	Experiment	Numerical	Experiment	Numerical	Experiment	Numerical
0.2	0.991	0.992	153.3	134.8	0.62	0.58
0.3	0.978	0.977	153.8	139.1	1.42	1.99
0.4	0.959	0.955	153.5	141.8	2.72	3.96
0.5	0.931	0.915	152.9	143.5	4.79	8.75

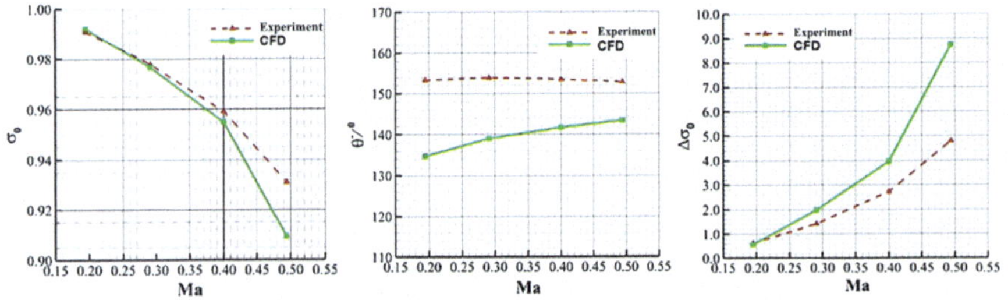

(a) Surface-averaged total pressure recovery coefficient σ_{av}

(b) Low-pressure region θ^-

(c) Steady-state distortion coefficient $\bar{\sigma}$

Fig. 7.13 Comparison diagram of steady-state distortion coefficient between numerical and experimental results for a 0.2*D* crescent-shaped distortion simulator

example, DDES numerical simulations are conducted based on the RANS flow field. Subsequently, extract the DDES numerical simulation results corresponding to the locations of steady-state and dynamic total pressure measurement points illustrated in Fig. 7.12, and calculate the steady-state and dynamic distortion coefficients. Figure 7.14 portrays an iso-surface plot of the *Q*-criterion of a transient flow field obtained from DDES numerical simulations, highlighting the robust analytical capability of DDES in capturing vortex structures effectively.

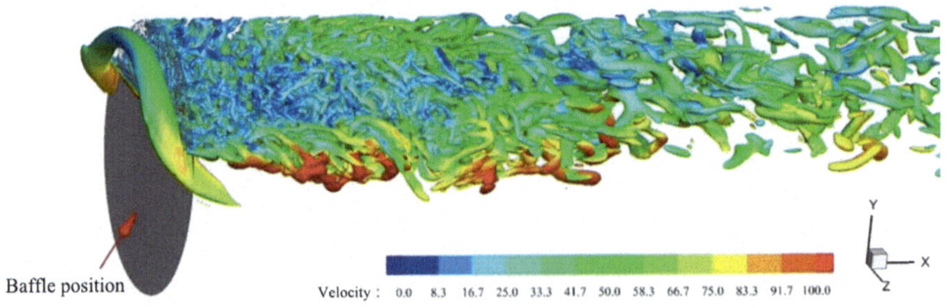

Fig. 7.14 Transient vortex structure diagram of a SD generator (based on the *Q*-criterion)

Table 7.8 Comparison of distortion coefficients for a SD generator with baffle depth of $0.2D$ between numerical and experimental results

	Steady-state distortion coefficient $\Delta\overline{\sigma}_0$		Dynamic distortion coefficient ε_{av}		$\Delta\overline{\sigma}_0/\varepsilon_{av}$	
	Experiment	Numerical	Experiment	Numerical	Experiment	Numerical
	0.62%	0.58%	0.495%	0.477%	1.25%	1.22%
Error (%)	6		4		2.4	

The results presented in Table 7.8 reveal the comparison between experimental and DDES computational outcomes for both the steady-state distortion coefficient $\Delta\overline{\sigma}_0$ and the mean dynamic distortion coefficient ε_{av}, along with their respective relative errors. Remarkably, the discrepancy between numerical and experimental values for the steady-state and dynamic distortion coefficients is merely 6% and 4%, respectively. Furthermore, the ratio of steady-state and dynamic distortion coefficients between experimental and numerical results demonstrates a minor difference of only 2.4%, with experimental results at 1.25 and numerical results at 1.22, which meet the demands of the engineering. It proves the accuracy of the DDES for such problem.

7.5 Analysis of the Impact of Flow Field Distortion Coefficients

The baffle depth stands as a pivotal factor, which directly affects the inlet flow rate of the distortion generator and exerts a significant influence on both steady-state and dynamic distortion coefficients. Consequently, this section undertakes an analysis of the impact patterns of varying baffle depths on distortion coefficients across three configurations: single-baffle, tandem double-baffle, and circumferential multi-baffle. Emphasis is placed on scrutinizing the influence of baffle depth variations on the ratio of steady-state to dynamic distortion coefficients. Furthermore, the section delves into assessing the influence of different configurations on distortion coefficients for identical baffle depths.

7.5.1 The Influence of Baffle Depth on Distortion Coefficients in Single-Baffle Distortion Generator

Table 7.9 provides the numerical simulation results for the steady-state, dynamic, and their ratio of distortion coefficients for SD generator with baffle depths $H = 0.2D, 0.3D, 0.4D$. As the baffle depth increases, the steady-state distortion coefficient increases from 0.58 to 3.35%, and the dynamic distortion coefficient increases from

0.48 to 2.11%, which indicates a gradual increase in both dynamic and steady-state distortion coefficients with increasing baffle depth.

How does an increase in the depth of the baffle affect the steady-state distortion coefficient? According to Eq. (7.1), the steady-state distortion coefficient is determined by the surface-averaged total pressure recovery coefficient σ_{av} and the circumferential total pressure recovery coefficient within the low-pressure zone σ_0. Figure 7.15 illustrates the distribution of the total pressure recovery coefficient along the circumference on the AIP. As the baffle depth increases, the range of lower total pressure recovery coefficients expands, indicating an enlargement of the low-pressure zone. Simultaneously, the surface-averaged total pressure recovery coefficient diminishes (Fig. 7.16), leading to an augmentation in the steady-state distortion coefficient.

Form Eq. (7.6), the value of dynamic distortion coefficient is upon the amplitude of total pressure pulsations on the AIP. Figures 7.17 and 7.18 present the distribution of the dynamic total pressure distortion coefficient at various azimuthal positions. Notably, the circumferential distribution of the dynamic distortion coefficient tends to exhibit a symmetrical pattern around the center of the low-pressure region ($\theta = 180°$). Moreover, this distribution is significantly influenced by the baffle depth

Table 7.9 Comparison of numerical simulation and experimental results for distortion coefficients of SD generator with three baffle depths

Baffle depth H	Steady-state distortion coefficient $\Delta\overline{\sigma}_0$ (%)	Dynamic distortion coefficient ε_{av} (%)	$\Delta\overline{\sigma}_0/\varepsilon_{av}$
0.2D	0.58	0.48	1.22
0.3D	1.79	1.13	1.58
0.4D	3.36	2.11	1.59

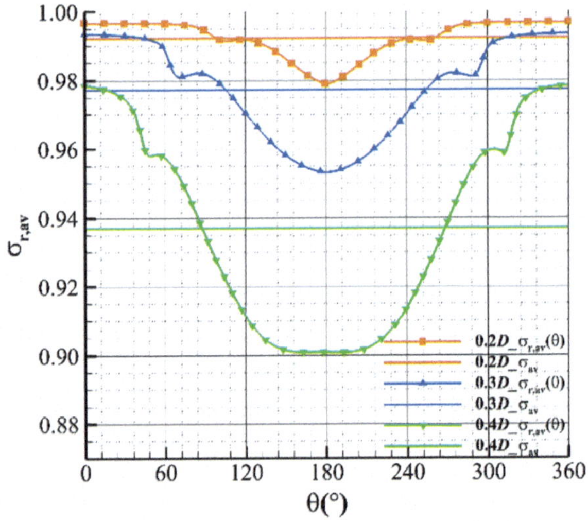

Fig. 7.15 The circumferential distribution of the $\sigma_{r,av}$ on the AIP

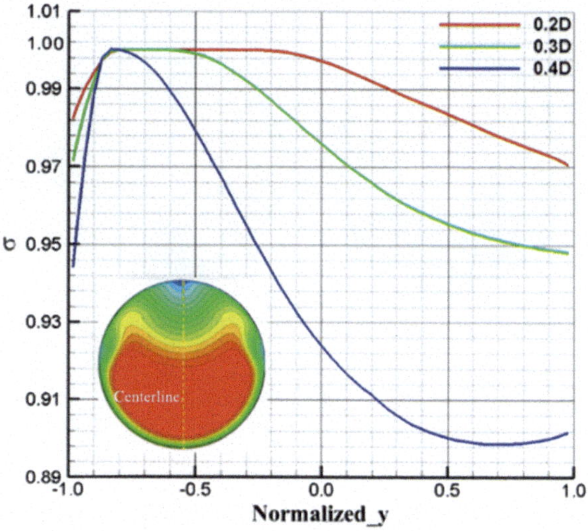

Fig. 7.16 The distribution of the σ along the centerline of the AIP

H. When the H is relatively small, the distribution typically manifests an inverted "V" shape. However, with increasing H, the baffle's impact on the mainstream flow becomes more pronounced, exacerbating turbulence pulsations. Consequently, there is a general augmentation in the dynamic distortion coefficient across all measurement points, accompanied by a downward shift in the location of maximum pulsation. At this juncture, the central region of the low-pressure zone ($\theta = 180°$) exhibits lower values, juxtaposed against higher values at the periphery, forming an "M" shape. This phenomenon arises from the complex airflow dynamics in the mixing zones where the baffles meet the tube wall, resulting in higher dynamic distortion coefficient values compared to the central low-pressure zone. Additionally, due to the increase in baffle depth, the position of maximum dynamic distortion coefficient gradually migrates away from the central low-pressure zone.

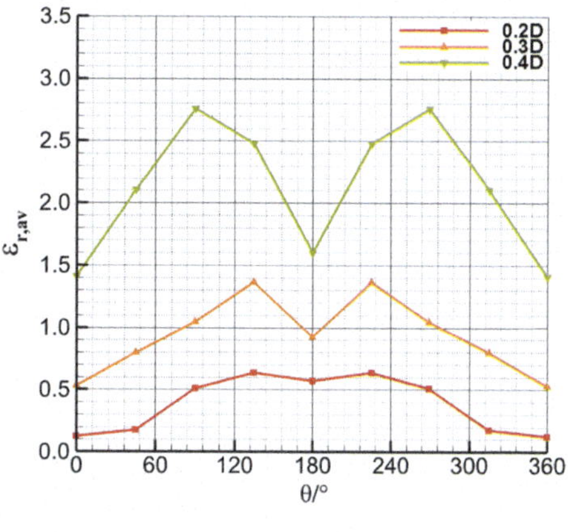

Fig. 7.17 The circumferential distribute on of the dynamic distortion coefficient $\varepsilon_{r,av}$ along the AIP for three different depths of baffle, resembling a crescent shape

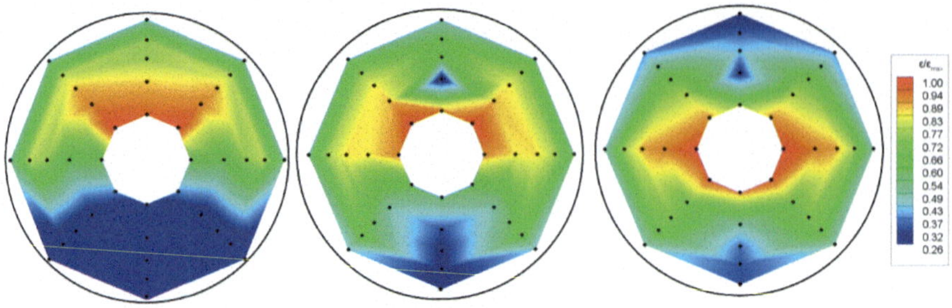

(a) 0.2*D* (b) 0.3*D* (c) 0.4*D*

Fig. 7.18 The distribution of the dynamic distortion coefficient ε along the AIP for three different depths of baffle, exhibiting a crescent shape

In the case of the crescent-shaped baffle configuration, on the AIP, the increase in baffle depth leads to concurrent increments in both steady-state and dynamic distortion coefficients. Consequently, the ratio of these coefficients, $\Delta\overline{\sigma}_0/\varepsilon_{av}$, exhibits a relatively minor range of variation. Specifically, as the baffle depth escalates from 0.2*D* to 0.4*D*, the ratio of steady-state to dynamic distortion coefficients merely rises from 1.22 to 1.59.

7.5.2 *The Influence of Tandem Double-Baffle Distortion Generator Depth and Axial Spacing on Distortion Coefficients*

In the context of the tandem double-baffle configuration, we conducted investigations across three distinct baffle depths. Table 7.10 outlines the steady-state distortion coefficients corresponding to these depths. As the baffle depth increases, the steady-state distortion coefficient increases from 0.86 to 2.47%. The reason for its increasement is the decrease in the surface-averaged total pressure recovery coefficient σ_{av} with a widening of the circumferential low-pressure zone, as Fig. 7.19 depicts.

Table 7.10 Table of steady-state distortion coefficients for three baffle depths in axial series baffles

Baffle depth	Surface-averaged total pressure recovery coefficient σ_{av}	Low-pressure region θ^- (°)	Steady-state distortion coefficient $\Delta\overline{\sigma}_0$ (%)
0.2*D*	0.988	148.5	0.86
0.3*D*	0.978	156.6	1.75
0.4*D*	0.954	175.5	2.47

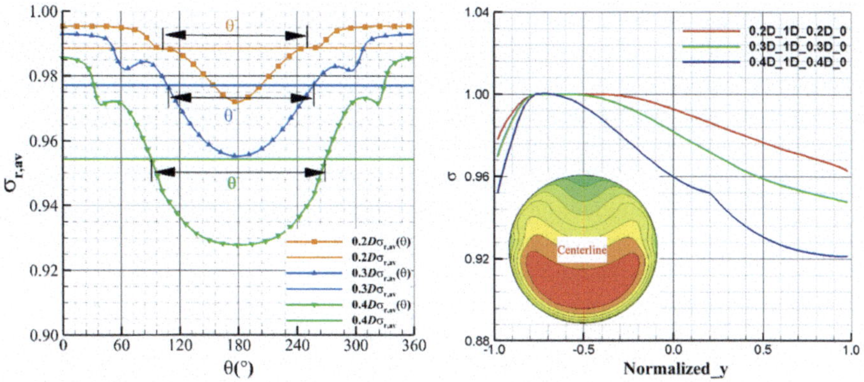

(a) Circumferential distribution of σ_{av}. (b) Distribution of σ along the central axis.

Fig. 7.19 Total pressure recovery coefficient distribution of the AIP along circumferential and central axis for three baffle depths in series baffles configuration

Utilizing the DDES numerical simulation outcomes, it's possible to delve into the impact of baffle depth on dynamic distortion coefficients as well as the ratio between steady-state and dynamic distortion coefficients. The numerical findings, as delineated in Table 7.11, illustrate that with the augmentation of baffle depth, the overall pressure pulsation range and amplitude at the AIP amplifies (Fig. 7.20), leading to a substantial rise in the dynamic distortion coefficient. This, in turn, engenders a reduction in the ratio between steady-state and dynamic distortion coefficients, declining from 1.72 at a depth of 0.2*D* to 1.01 at a depth of 0.4*D*.

Table 7.11 Dynamic distortion coefficients for different baffle depths in tandem double-baffle configurations

Baffle depth	Dynamic distortion coefficient ε_{av} (%)	$\Delta\overline{\sigma}_0/\varepsilon_{av}$
0.2*D*	0.50	1.72
0.3*D*	1.04	1.68
0.4*D*	2.44	1.01

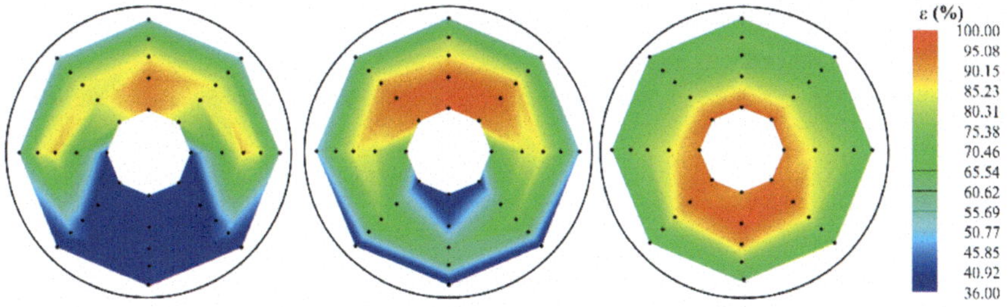

(*a*) **0.2*D*** series baffles (b) **0.3*D*** series baffles (c) **0.3*D*** series baffles

Fig. 7.20 Tandem double-baffle dynamic distortion coefficient ε at the AIP

To investigate the influence of axial spacing ΔL on the axial series combined baffle configuration, numerical simulations were conducted with $\Delta L = 0.1D, 0.5D$, at the circumferential position $\alpha = 0°$, baffle depth $H = 0.2D$ and incoming flow Mach number $Ma = 0.2$. Regardless of the axial position variation, the AIP remained consistently positioned 3D downstream from the baffle. The numerical results, detailed in Table 7.12, depict the steady-state distortion coefficient for various axial spacings. With a reduction in axial spacing from $1.0D$ to $0.5D$, the steady-state distortion coefficient decreases from 0.86 to 0.51%, closely aligning with the outcomes derived from the traditional chordal method ($\Delta\overline{\sigma}_0 = 0.58\%$).

From the flow field perspective in the meridional plane, as illustrated in Fig. 7.21, for a crescent-shaped baffle with a depth of $H = 0.2D$, the axial length of the recirculation zone behind the baffle is approximately $1D$. In the configuration of tandem double-baffles, when the axial spacing is only $0.5D$, the downstream baffle is positioned precisely within the recirculation zone of the upstream baffle. Although the downstream baffle divides the recirculation zone, its impact on the effective flow area can be largely disregarded. From the standpoint of steady-state distortion coefficient, in the tandem double-baffles configuration, when the spacing between the two baffles is smaller than the length of the recirculation zone behind the baffle, the effect of the downstream baffle can be ignored.

Based on the DDES numerical simulation results, the dynamic distortion coefficient of the tandem double-baffle configuration decreases from 0.50 to 0.33% when the axial spacing reduces from $1.0D$ to $0.5D$ as shown in Table 7.13. Comparing Fig. 7.22b to Fig. 7.18a, the distribution of the dynamic distortion coefficient at an axial spacing of $0.5D$ resembles that of the series baffle configuration. This suggests that the influence of the downstream baffle not only weakens the effect on the steady-state distortion coefficient but also diminishes its impact on total pressure pulsations. Overall, when the axial spacing decreases, both steady-state and dynamic distortion coefficients decrease, but the reduction of the former is more pronounced. Consequently, the ratio between the two decreases. Thus, it is evident that the spacing between baffles in the axial series configuration shouldn't be less than the length of the recirculation zone behind the upstream baffle when increasing either the steady-state or dynamic distortion coefficient.

Table 7.12 Dynamic distortion coefficient for different axial spacings in tandem double-baffle configuration

Baffle configuration	Surface-averaged total pressure recovery coefficient σ_{av}	Low-pressure region θ^- (°)	Steady-state distortion coefficient $\Delta\overline{\sigma}_0$ (%)
Axial series spacing $1.0D$	0.988	148.5	0.86
Axial series spacing $0.5D$	0.992	122.9	0.51

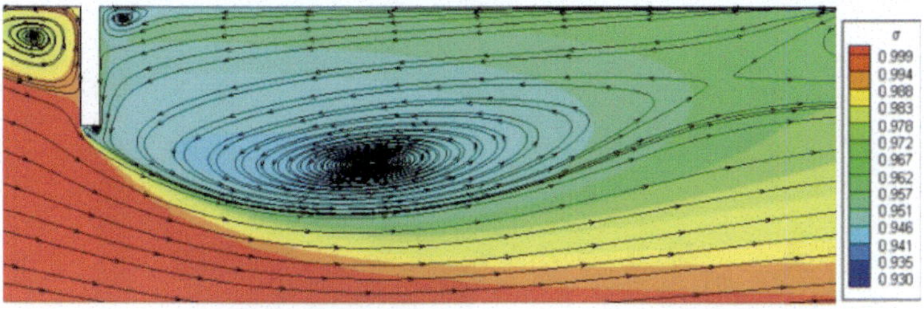

(a) Single-baffle distortion generator.

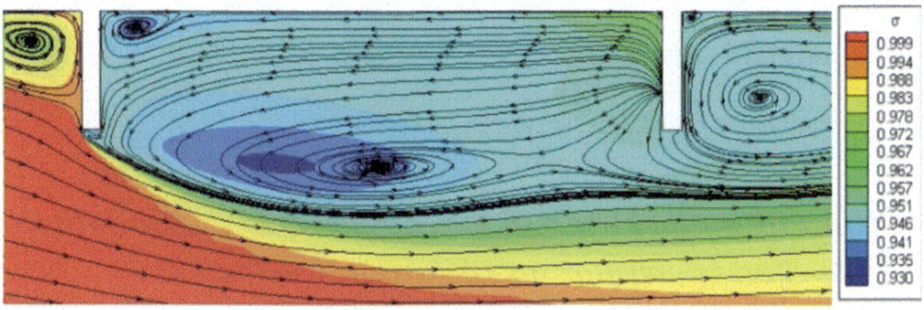

(b) Tandem double-baffle distortion generator (axial spacing $1.0D$).

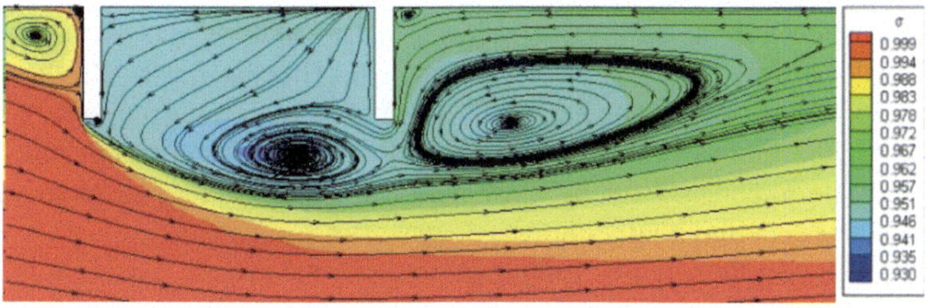

(c) Tandem double-baffle distortion generator (axial spacing $0.5D$)

Fig. 7.21 Total pressure recovery coefficient and streamline distribution in the meridional plane (tandem double-baffle distortion generator)

Table 7.13 Dynamic distortion coefficient for different axial spacings in tandem double-baffle distortion generators

Baffle configuration	Dynamic distortion coefficient ε_{av} (%)	$\Delta\overline{\sigma}_0/\varepsilon_{av}$
Axial spacing 1.0D	0.50	1.72
Axial spacing 0.5D	0.33	1.56

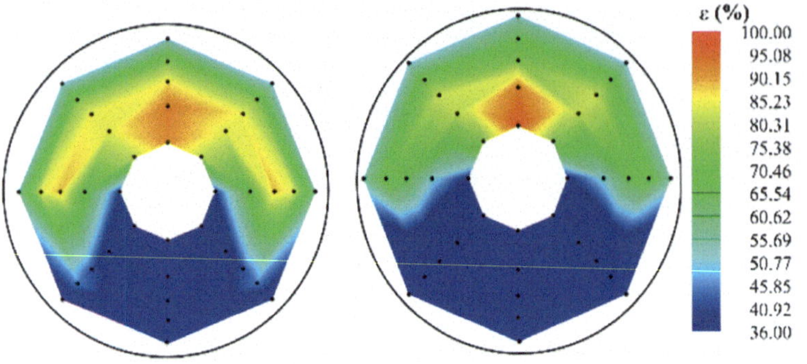

(a) Axial spacing $1D$. (b) Axial spacing $0.5D$.

Fig. 7.22 The distribution of the dynamic distortion coefficient ε at the AIP

7.5.3 Effect of Baffle Depth in Circumferential Multi-baffle Distortion Generator on Distortion Coefficient

The characteristic feature of the circumferential multi-baffle distortion generator lies in the presence of multiple axially symmetric low-pressure zones, the number of which correlates with the number of baffles. This subsection focus on the scenario where there are four baffles, baffle depths $H = 0.2D, 0.3D, 0.4D$, and the inflow Mach number $Ma = 0.2$. Figure 7.23 is the distribution of the total pressure recovery coefficient in the vicinity of the baffle. The results obtained from steady-state calculations indicate a tendency for the flow field of multiple baffles to progressively become more circumferentially uniform downstream.

As the baffle depth increases from $0.2D$ to $0.4D$, the average surface total pressure recovery coefficient σ_{av} decreases from 0.988 to 0.981 (refer to Table 7.14) with negligible impact. Regarding the circumferential parallel configuration, the predominant influence on the range of the low-pressure zone is attributed to the number of baffles, as depicted in Fig. 7.24. Increasing the baffle depth exhibits minimal effect on the extent of the low-pressure zone θ^{-}. Consequently, the capability of baffle depth to alter the steady-state distortion coefficient is constrained.

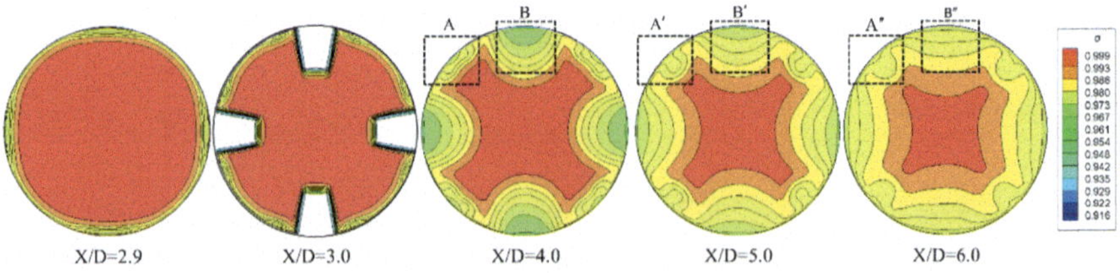

Fig. 7.23 The diagram of the total pressure recovery coefficient at typical cross-sections before, beneath, and after the plugboard insertion

Table 7.14 Table of steady-state distortion coefficient results for circumferential multi-baffle distortion with three baffle depths

Baffle depth	Surface-averaged total pressure recovery coefficient σ_{av}	Low-pressure region θ^- (°)	Steady-state distortion coefficient $\Delta\overline{\sigma}_0$ (%)
0.2*D*	0.988	179.9	0.12
0.3*D*	0.984	180.0	0.22
0.4*D*	0.981	180.0	0.24

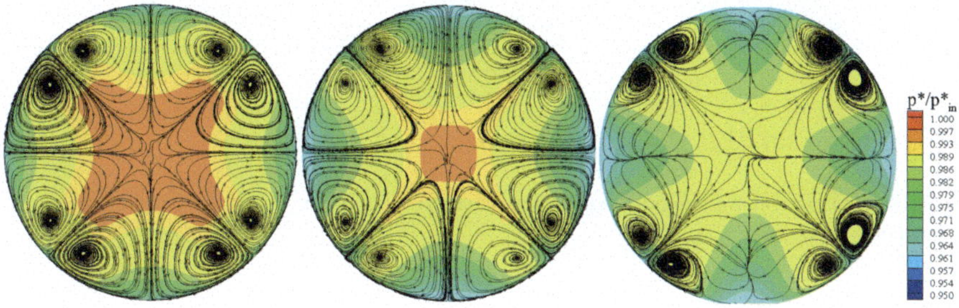

(a) $h = 0.2D$ (b) $h = 0.3D$ (c) $h = 0.4D$

Fig. 7.24 Distribution of AIP static pressure and secondary flow streamlines for three baffle depths

Table 7.15 Dynamic distortion coefficient for different baffle depths in CMD configuration

Baffle depth	Dynamic distortion coefficient ε_{av} (%)	$\Delta\overline{\sigma}_0/\varepsilon_{av}$
0.2*D*	0.53	0.22
0.3*D*	0.66	0.33
0.4*D*	0.67	0.36

Then undertake an analysis of the impact of circumferential multizone combination configurations on the dynamic distortion coefficient by leveraging the results of DDES simulations. Owing to the circumferential symmetry of this configuration, employ axisymmetric boundary treatment during computation to optimize computational efficiency. The dynamic distortion coefficient outcomes are in Table 7.15. As the baffle depth increases from 0.2*D* to 0.4*D*, the dynamic distortion coefficient first gradually increases (from 0.53 to 0.66%), followed by a period of relative stability (0.66–0.67%). From the distribution of the dynamic distortion coefficient on the AIP in Fig. 7.25, the circumferential multizone combination dynamic distortion coefficient is uniformly dispersed, which echoes the effect of baffle depth on the steady-state distortion coefficient for this configuration, characterized by the absence of pronounced regions of heightened fluctuation. In summation, the capacity of this configuration to modulate the ratio of steady-state and dynamic distortion coefficients ($\Delta\overline{\sigma}_0/\varepsilon_{av}$) through adjustments in baffle depth is constrained.

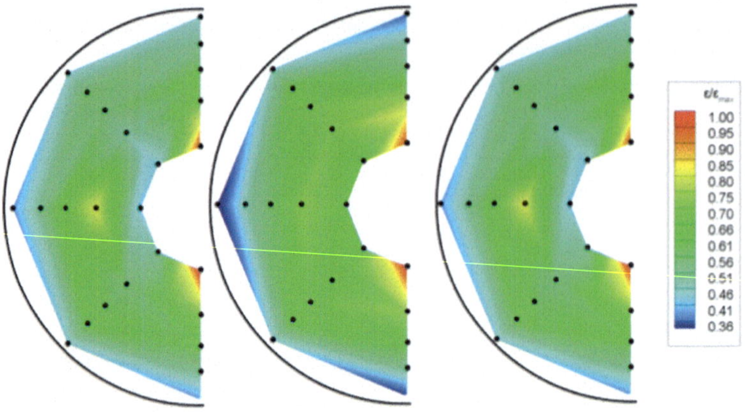

(a) $0.2D$ (b) $0.3D$ (c) $0.4D$

Fig. 7.25 Distribution of dynamic distortion coefficient ε on AIP for CMD configuration with three baffle depths (symmetrical)

References

1. Horvath N R. Inlet Vortex Formation Under Crosswind Conditions. America, Massachusetts: Worcester Polytechnic Institute, 2013.
2. Zheng R, Li W, Chang J, et al. Suppression of Interaction Between Shock Wave and Boundary Layer for "X" Hypersonic Inverted Inlet. Journal of Propulsion Technology, 2014, 35(09): 1153–1161.
3. McLelland, Grant; MacManus, David G.; Zachos, Pavlos K.; Gil-Prieto, Daniel; Migliorini, Matteo. Influence of Upstream Total Pressure Profiles on S-Duct Intake Flow Distortion. Journal of Propulsion and Power, 2020, 36(3), 346–356.
4. Liu L, Chen F, Song Y, et al. Numerical Simulation of Effects of Active Flow Control on the Performance of a Half Flush-Mounted S-Shaped Inlet. Journal of Propulsion Technology, 2014, 35(09): 1168–1173.
5. Liu, D X, Ye, P L, Hu, J, et al. Stability Design and Evaluation Technology for Aircraft Gas Turbine Engines. Beijing: Aviation Industry Press, 2004.
6. Rademakers R. P. M., Bindl S., Niehuis R. Effects of Flow Distortions as They Occur in S-Duct Inlets on the Performance and Stability of a Jet Engine. Journal of Engineering for Gas Turbines and Power, 2016, 138(2): 1–10.
7. Du J, Han W. Calculation of the average turbulence in the inlet pressure distortion experiment. Gas Turbine Experiment and Research, 2019(3): 53–57.
8. Yang Q, Ye W, Lu D. A Selection of Testers for Aero-engine Stability Assessment. Gas Turbine Experiment and Research, 2001, 14(02): 16–21.
9. Aviation Science and Technology Commission. Guidelines for the Assessment of Inlet Total Pressure Distortion in Aircraft Turboshaft and Turbofan Engines, GJB/Z 64A-2004. Beijing: China Standards Press, 2004: 1–13.
10. Lee K, Lee B, Kang S. Inlet Distortion Test with Gas Turbine Engine in the Altitude Engine Test Facility. Chicago: 27th AIAA Aerodynamic Measurement Technology and Ground Testing Conference. 2010.
11. Sang Z, Jiang Y, Kong W. Experimental Study on the Inlet Total Pressure Distortion of a Turbojet Engine. Journal of Aerospace Power, 2000, 15(04): 423–426.
12. Jiang Y, Li J, Zhang Q. An Inlet Pressure Distortion Measurement and Control System for Jet Propulsion System. Metrology & Measurement Technology, 2002, 22(06): 16–20.

13. Gan T, Wang R, Zhang J. Numerical Simulation of Inlet Distortion with Interceptor with Different Turbulence Models. Journal of Propulsion Technology, 2014, 35(07): 891–896.
14. Zhang X, Li J, Song G. Numerical simulation on flat baffle inlet distortion of axial compressor. Journal of Aerospace Power, 2019, 34(05): 1153–1165.
15. Liu Y, Yan H, Liu Y, et al. Numerical study of corner separation in a linear compressor cascade using various turbulence models. Chinese Journal of Aeronautics, 2016, 29(3): 639–652.
16. Jiang J, Zhao H, Li J. Prediction and test verification of inlet distortion based on flight test data. Journal of Aerospace Power, 2020, 35(08): 1706–1715.

The manufacturer's authorised representative in the EU is Springer Nature Customer Service Centre GmbH, Europaplatz 3, 69115 Heidelberg, Germany. If you have any concerns regarding our products, please contact ProductSafety@springernature.com

Printed and bound by CPI Group (UK) Ltd, Croydon, CR0 4YY
01/10/2026
02225937-0002